Chenkun Qi · Feng Gao · Yan Hu

Control of Hardware-In-The-Loop Space Contact Simulation Robot

Springer

Chenkun Qi
State Key Laboratory of Mechanical System and Vibration
School of Mechanical Engineering
Shanghai Jiao Tong University
Shanghai, China

Feng Gao
State Key Laboratory of Mechanical System and Vibration
School of Mechanical Engineering
Shanghai Jiao Tong University
Shanghai, China

Yan Hu
State Key Laboratory of Mechanical System and Vibration
School of Mechanical Engineering
Shanghai Jiao Tong University
Shanghai, China

ISBN 978-981-95-1171-6 ISBN 978-981-95-1172-3 (eBook)
https://doi.org/10.1007/978-981-95-1172-3

This Springer imprint is published by the registered company Springer Nature Singapore Pte Ltd.
The registered company address is: 152 Beach Road, #21-01/04 Gateway East, Singapore 189721, Singapore

Preface

The book focuses on control methods of hardware-in-the-loop simulation robot on the ground for simulating contact dynamics of spacecraft and space manipulator in space. Both principles and engineering practice have been addressed. This is achieved by providing an in-depth study on several major topics such as control system structure and modeling, simulation distortion analysis, distortion compensation control methods, simulation and experimental verifications, and applications. The comprehensive and systematic treatment of practical issues in control of hardware-in-the-loop space contact simulation robot is one of the major features of the book, which is particularly suited for readers who are interested to learn practical solutions in control of hardware-in-the-loop space contact simulation robot. The book can benefit researchers, engineers, and graduate students in the fields of aerospace engineering, mechanical engineering and control engineering, etc.

Authors are grateful to students and colleagues in our research group for their support and contributions, and also would like to thank the National Natural Science Foundation of China (Grant Nos. 51927809 and 51975351).

Shanghai, China

Chenkun Qi
Feng Gao
Yan Hu

Competing Interests The authors have no competing interests to declare that are relevant to the content of this manuscript.

Contents

Abbreviations

0-g	Zero-Gravity
1-g	One-Gravity
ADC	Actuation Delay Compensation
AMFC	Actuation and Mechanism Force Compensation
CDC	Calculation Delay Compensation
CM	Capture Mechanism
CoR	Coefficient of Restitution
DM	Docking Mechanism
DOF	Degree-of-Freedom
DRD	Dynamic Response Delay
DRDC	Dynamic Response Delay Compensation
EPOS	European Proximity Operations Simulator
FDC	Force Delay Compensation
FKM	Forward Kinematic Model
FMD	Force Measurement Delay
FMDC	Force Measurement Delay Compensation
HIL	Hardware-In-the-Loop
HMI	Human-Machine-Interface
IKM	Inverse Kinematics Model
LAC	Learning-based Actuation Compensator
LFC	Learning-based Force Compensator
LSC	Learning-based Structure Compensator
MDC	Mechanism Delay Compensation
MIMO	Multi-Input Multi-Output
PRS	Prismatic-Revolute-Spherical
PSS	Prismatic-Spherical-Spherical
RDOTS	Rendezvous and Docking Operation Testing System
RMS	Remote Manipulator System
SDC	Structural Dynamics Compensation

SDTS	Six-Degree-of-Freedom Test System
SHM	Simple Harmonic Motion
SPDM	Special Purpose Dexterous Manipulator
SRMS	Shuttle Remote Manipulator System
STVF	SPDM Task Verification Facility

List of Figures

List of Tables

Chapter 1
Introduction

1.1 Background

With the development of advanced space technologies, the complicated space missions become possible. Space rendezvous and docking technologies promote the space on-orbit services, such as transportation, maintenance and upgrade. The first typical example is the spacecraft docking, where two spacecraft gradually approach and finally dock through the docking mechanisms (DM) (Ding et al. 2020). The docking process includes a series of contacts between an active DM and a passive DM. The contact dynamics of docking mechanisms in space is very important for the success of the space docking mission. The contact process of space docking is very complex due to the complicated contact surface and contact characteristics. To design and test the docking system, and obtain the docking parameters (e.g., initial relative posture, initial relative velocity and maximum contact force), the contact dynamics of docking mechanisms should be simulated on the ground before it is launched (Xu et al. 2011; Flores-Abad et al. 2014).

The second typical example is the space manipulator on-orbit operation. The space manipulator is very useful for on-orbit services, such as the space station construction and maintenance. For the target capture of space manipulator, the space manipulator with an end-effector gradually approach the target spacecraft and finally capture the target adaptor on the target spacecraft. The end-effector and target adaptor together form the capture mechanisms (CM). The contact dynamics of space capture is also very complex due to the complicated contact surface and contact characteristics. The simulation of the target capture process of space manipulator is of great significance for the manipulator, capture mechanism and capture task verification (Aghili 2019).

The contact process of docking or capture mechanisms happens in the zero-gravity (0-g) environment in space. However, on the ground the simulation of space contact is in the one-gravity (1-g) environment. The difference of the gravity environment makes the contact dynamics simulation of docking or capture mechanisms very

C. Qi et al., *Control of Hardware-In-The-Loop Space Contact Simulation Robot*,
https://doi.org/10.1007/978-981-95-1172-3_1

difficult (Xu et al. 2011; Flores-Abad et al. 2014). The simulation fidelity and system stability are very important for the space contact simulator.

For the simulation of space contact on the ground, there are three classes of approaches: numerical simulation, hardware simulation and hardware-in-the-loop (HIL) simulation. For the numerical simulation, only the computer software is used to simulate the contact dynamics and the motion of spacecraft in the 0-g environment (Ma 1995; Ma and Wang 2007; Ma et al. 1997; Yaskevich 2014). Thus, it is convenient and flexible to change the spacecraft model. The simulation of spacecraft motion after a force is accurate enough, however it is not easy to simulate the contact process accurately due to the inaccurate contact model, e.g., inaccurate material properties and inaccurate contact stiffness. Therefore, contact experiments on real hardware have to be used to get more accurate results, which can also calibrate the numerical simulation model.

For the physical simulation, the 0-g environment is emulated using specially designed platforms, such as air-bearing supported floating platform (Uyama et al. 2012; Romano et al. 2007; Schwartz et al. 2003; Ma and Yang 2006) and suspension system (Riley et al. 1966; Hatch et al. 1967). The real docking mechanisms with the real spacecraft or equivalent spacecraft model are put in the 0-g emulation environment to simulate the motion and contact dynamics of spacecraft in space. Thus the simulation fidelity could be better. However, the 0-g emulation system is very complicated and the spacecraft model is inconvenient to change.

For the HIL simulation (also called hybrid simulation), the numerical simulation software is used to compute the motion of spacecraft in the zero-gravity environment, and the robotic motion simulator is used to reproduce the motion of the spacecraft, and the real docking hardware are used to generate the real contact process. The contact process happens on the real docking hardware. Because the HIL simulation integrates the fidelity of physical simulation and the flexibility of numerical simulation, it is a very useful approach to simulate the complicated contact process in space (Xu et al. 2011; Ma et al. 2012; Rems et al. 2021).

In general, a HIL simulator for space docking consists of a motion simulator, real docking mechanisms, a contact force and torque measurement system, and a numerical simulation system for the dynamics of the spacecraft. The motion simulator includes a lower motion simulator (i.e., a lower platform) and an upper platform (i.e., a fixed or movable platform). The real docking mechanisms include an active docking mechanism and a passive docking mechanism. The contact force and torque measurement system includes a contact force and torque sensor and a signal collection and processing system. The numerical simulation system for the dynamics of spacecraft includes two mathematical models for the dynamics of an active spacecraft and a passive spacecraft and a real-time computing software module. By synthesizing the physical hardware (i.e., motion simulator, real docking mechanisms, and contact force and torque measurement system) and the numerical simulation of the active spacecraft and the passive spacecraft, the HIL simulator can reproduce the real space contact process on the ground without requiring a real zero-gravity facility. Moreover, the HIL simulator is flexible to change flying objects (i.e., spacecraft) through the software settings.

For the HIL contact simulators, the robots are often used to reproduce the relative motion of spacecraft in space. Among them, serial mechanism robots are widely used as motion simulators, such as the hydraulic manipulator for special purpose dexterous manipulator (SPDM) task verification facility (STVF) at the Canadian space agency (Piedboeuf et al. 1999; Ma et al. 2004; Martin et al. 2005; Zhu et al. 2002), two industrial robots for European proximity operations simulator (EPOS) at the German aerospace center (Ma et al. 2012; Boge and Ma 2011; Zebenay et al. 2015), the ground experimental system of remote manipulator system (RMS) at the Japan aerospace exploration agency (Sawada et al. 2007), the ground experiment system of free-floating space robot for capturing space target at the Shenzhen space technology center (Xu et al. 2007), and the HIL simulator of space manipulator at the Tsinghua university (Mou et al. 2018). The Stewart parallel mechanism robots have also been used as motion simulators, such as the six-degree-of-freedom test system (SDTS) at the NASA Johnson space center (Mitchell et al. 2008), the ground-based hydraulic simulator of shuttle remote manipulator system (SRMS) at the McDonnell Douglas aerospace (Ananthakrishnan et al. 1996), space station docking mechanism dynamic testing system at the Marshall space flight center (Howsman and Glaese 1988), spacecraft docking and space manipulator capture simulation platform at the Harbin institute of technology (Huang et al. 2005; Chang et al. 2007; Yu et al. 2018a, b), the rendezvous and docking operation testing system (RDOTS) at the Japanese space agency (Lange et al. 2002), the collision HIL motion simulator at the Tohoku university (Takahashi et al. 2008; Osaki et al. 2010; Abiko et al. 2014).

Due to the structure difference between the space docking system and the ground HIL simulation system, it is very difficult to achieve high accuracy of HIL simulation. To ensure experimental accuracy, the design of the HIL simulator is very important. Different from a noncontact rendezvous simulator, a highly accurate HIL simulation of space contact requires that the motion simulator should have high positioning accuracy, high bandwidth frequency response and low time delay.

1.2 Control Methods of HIL Simulation Robot of Space Contact

An ideal HIL simulator requires that all subsystems have no time delays and errors. However, it is difficult to satisfy these technical requirements using available mechanical and control systems. For a real HIL simulator, there are always time delays or non-ideal dynamics due to the subsystem performance limitations.

For example, due to the limited bandwidth of the force/torque sensor, signal acquisition and noise filtering, the force measurement system has a force measurement delay. The numerical simulation of dynamics model of service spacecraft (or space manipulator) and target spacecraft has a numerical calculation delay. Due to the limited bandwidth of the actuation system and external inertia, the lower motion simulator generates the dynamic response delay during tracking the motion command.

Because the mechanical structures of the lower and upper motion simulator have mass/inertia and damping, and their structural stiffness is limited, the mechanical systems generate structural deformation dynamics.

One challenging issue of the space docking HIL simulator is that these time delays or non-ideal dynamics in the system loops lead to the simulation distortion and even system instability (Sazawal et al. 2019). The HIL simulation distortion often shows the velocity divergence or convergence. This means that the coefficient of restitution (CoR) of velocity (i.e., the ratio of the rebound velocity after the contact and the approach velocity before the contact) is not equal to what occurs in space. The HIL simulation of space docking is often divergent due to the phase delay in the system loops (De Stefano et al. 2020). The convergence distortion can also occur due to dissipative factors, e.g., the structural deformation dynamics of the upper motion simulator. Although some HIL simulation testbeds for space docking have been established, the distortion and instability problem is not well studied. One way is to use a passive compliant device to solve the instability problem. The stability region with compliance parameters is studied in Sazawal et al. (2019). However, the passive compliant device can not handle the convergence distortion problem. Moreover, it is not easy to guarantee the simulation accuracy. To better improve the simulation accuracy and guarantee the system stability, the distortion compensation control is a necessary and effective way because it can actively compensate for the time delays and non-ideal dynamics.

Some compensation control algorithms have been studied to solve the HIL simulation distortion problem. According to time delays or non-ideal dynamics to be compensated, the compensator can be summarized as follows. Force measurement delay (Osaki et al. 2010; Qi et al. 2016; Chang et al. 2007; Qi et al. 2017a, c; Wang et al. 2019), dynamic response delay (Ananthakrishnan et al. 1996; Qi et al. 2017a, b, c; Wang et al. 2019), servo system delay (Osaki et al. 2010; Abiko et al. 2014; Chang et al. 2007), mechanical inertia delay (Shimoji et al. 1991), numerical calculation delay (Ananthakrishnan et al. 1996; Osaki et al. 2010; Shimoji et al. 1991; Chang et al. 2007), communication delay (Osaki et al. 2010), data filtering delay (Abiko et al. 2014), force sensor deadband (Abiko et al. 2014), discrete integration (De Stefano et al. 2020), have been considered in the distortion compensator. Authors' studies consider the structural dynamics of the upper platform (Qi et al. 2018b; Qi et al. 2021) and the structural dynamics of the lower motion simulator (i.e., the lower platform) (Gao et al. 2016; Qi et al. 2022; Qi et al. 2023a).

According to the output variable, the compensator can be classified into force compensator, position compensator and composite compensator.

The force compensator outputs a force signal by adding a compensation force on the measured force of the contact. For example, a phase lead force compensator is designed to compensate for the time delay of the measured contact force (Shimoji et al. 1991; Osaki et al. 2010; Qi et al. 2016; Qi et al. 2018a). The virtual force compensation (Zebenay et al. 2015, 2013; Osaki et al. 2010; Abiko et al. 2014) adds a virtual force onto the measured force to compensate for one dead time. To compensate for the dynamic response delay of the motion simulator, a force compensator is designed based on the contact stiffness and damping identification (Qi et al. 2017a, b; Yu et al. 2018b; Wang et al. 2019). The structural deformation of the motion simulator can also be compensated by the force compensator (Qi et al. 2018b; Gao et al. 2016).

The position compensator outputs a position signal by correcting the position command of the motion simulator. For example, the dynamic response delay of the motion simulator is compensated by a position command compensator using a low-order model (Qi et al. 2017c). Feed-forward position compensation (Ananthakrishnan et al. 1996; Chang et al. 2007) is used to correct the command position of motion simulator and makes its actual position approximate the desired position of spacecraft.

Some composite compensators output the force and velocity signals. In (De Stefano et al. 2020), the force and velocity compensators are designed using the passivity theory to compensate for the discrete-time integration and the time delay in the computational loop. Some composite compensators output the force and position signals. In (Qi et al. 2022), the delays from the force, calculation, actuation and mechanism in the system loop are compensated in a distributed compensation framework with both force compensator and position compensators. In (Yu et al. 2022), smith predictor and fuzzy control are used to compensate for the force measurement delay and the dynamic response delay of the motion simulator respectively.

Depending on the way the compensator is designed, the compensators can be classified into model-based compensator and data-based compensator.

When the delay model is known, the distortion compensator can be designed based on the model. For example, the phase lead time of the force compensator can be set as the total delay time of the HIL system loops (Shimoji et al. 1991; Osaki et al. 2010; Qi et al. 2016; Qi et al. 2017a). When the contact stiffness is also known, a first-order compensator can be designed to achieve the desired CoR (Osaki et al. 2010; Abiko et al. 2014).

When the delay model is unknown, the compensator can be designed based on the experimental data. For example, the compensator can be tuned using the trial-and-error method (Chang et al. 2007; Zebenay et al. 2015) or offline optimization method based on experimental data (Ananthakrishnan et al. 1996). A learning-based neural network compensator is proposed to compensate for phase delays and structural dynamics of the simulator in (Qi et al. 2023b).

1.3 Contributions and Organization of the Book

The book focuses on control methods of the HIL simulation robot for simulating spacecraft contact dynamics in space. The main contributions of the book include that the modeling and distortion analysis of the HIL simulation robot are presented, and systematic distortion compensation control methods are proposed.

The rest of the book is organized as follows. Chapter 2 introduces the robot system for HIL simulation of space contact dynamics. The compensation control of force measurement delay, dynamic response delay of lower motion simulator, structural dynamics of lower motion simulator, structural dynamics of upper platform are introduced in Chaps. 3, 4, 5 And 6 respectively. Chapters 7 and 8 provide the lumped and distributed distortion compensation control respectively. A learning based distortion compensation control is given in Chap. 9. The last chapter gives some conclusions.

References

Abiko, S., Satake, Y., Jiang, X., Tsujita, T., & Uchiyama, M. (2014). Time delay compensation based on coefficient of restitution for collision hybrid motion simulator. Advanced Robotics, 28(17), 1177–1188.

Aghili, F. (2019). Robust impedance-matching of manipulators interacting with uncertain environments: application to task verification of the space station's dexterous manipulator. IEEE/ASME Transactions on Mechatronics, 24(4), 1565–1576.

Ananthakrishnan, S., Teders, R., & Alder, K. (1996). Role of estimation in real-time contact dynamics enhancement of space station engineering facility. IEEE Robotics and Automation Magazine, 3(3), 20–28.

Boge, T., & Ma, O. (2011). Using advanced industrial robotics for spacecraft rendezvous and docking simulation. Proceedings of the 2011 IEEE International Conference on Robotics and Automation, 09–13 May, 2011, Shanghai, China.

Chang, T., Cong, D., Ye, Z., & Han, J. (2007). Time problems in HIL simulation for on-orbit docking and compensation. 2nd IEEE Conference on Industrial Electronics and Applications, 23–25 May, 2007, Harbin, China.

De Stefano, M., Balachandran, R., & Secchi, C. (2020). A passivity-based approach for simulating satellite dynamics with robots: discrete-time integration and time-delay compensation. IEEE Transactions on Robotics, 36(1), 189–203.

Ding, X. L., Wang, Y. C., Wang, Y. B., & Xu, K. (2020). A review of structures, verification, and calibration technologies of space robotic systems for on-orbit servicing. Science China Technological Sciences, 64, 1–19.

Flores-Abad, A., Ma, O., Pham, K., & Ulrich, S. (2014). A review of space robotics technologies for on-orbit servicing. Progress in Aerospace Sciences, 68, 1–26.

Gao, F., Qi, C. K., Ren, A. Y., Zhao, X. C., et al. (2016). Hardware-in-the-loop simulation for the contact dynamic process of flying objects in space. Science China Technological Science, 59(8), 1167–1175.

Hatch, H. G., Pennington, J. E., & Cobb, J. B. (1967). Dynamic simulation of lunar module docking with Apollo command module in lunar orbit. NASA TN D-3972, p. 1–26.

Howsman, T. G., & Glaese, J. (1988). Space station docking mechanism dynamic testing. 15th Space Simulation Conference: Support the Highway to Space Through Testing, Goddard Space Flight Center, NASA, Jan 01, 1988, pp. 168–175.

Huang, Q. T., Jiang, H. Z., Zhang, S. Y., and Han, J. W. (2005). Spacecraft docking simulation using hardware-in-the-loop simulator with Stewart platform. Chinese Journal of Mechanical Engineering, 18(3), 415–418.

Lange, C., Martin, E., Piedboeuf, J.-C., & Kövecses, J. (2002). Towards Docking Emulation Using Hardware in the Loop Simulation with Parallel Platform. Proceedings of the Workshop on Fundamental Issues and Future Directions for Parallel Mechanisms and Manipulators, 3–4 October, 2002, Quebec, Canada.

Ma, O., & Wang, J. (2007). Model order reduction for impact-contact dynamics simulations of flexible manipulators. Robotica, 25, 397–407.

Ma, O., Buhariwala, K., Roger, N., MacLean, J., & Carr, R. (1997). MDSF - a generic development and simulation facility for flexible, complex robotic systems. Robotica, 15(01), 49–62.

Ma, O., Wang, J., Misra, S., & Liu, M. (2004). On the validation of SPDM task verification facility. Journal of Robotic Systems, 21(5), 219–235.

Ma, O., Flores-Abad, A., & Boge, T. (2012). Use of industrial robots for hardware-in-the-loop simulation of satellite rendezvous and docking. Acta Astronautica, 81, 335–347.

Ma, O., & Yang, G. (2006). Validation of a satellite docking simulator using the SOSS experimental testbed. Proceedings of the 2006 IEEE/RSJ International Conference on Intelligent Robots and Systems, 9–15 October, 2006, Beijing, China.

Ma, O. (1995). Contact dynamics modelling for the simulation of the space station manipulators handling payloads. Proceedings of the 1995 IEEE International Conference on Robotics and Automation, 21–27 May, 1995, Nagoya, Japan.

Martin, É., Doyon, M., Gonthier, Y., & Lange, C. (2005). Validation process of the STVF hardware-in-the-loop simulation facility. Proceedings of the 8th International Symposium on Artificial Intelligence, Robotics and Automation in Space - iSAIRAS, 5–8 September, 2005, Munich, Germany.

Mitchell, J. D., Cryan, S. P., Baker, K., Martin, T., Goode, R., Key, K. W., Manning, T., & Chien, C. H. (2008). Integrated docking simulation and testing with the Johnson space center six-degree-of-freedom dynamic test system. Space Technology and Applications International Forum—STAIF 2008 (AIP Conference Proceedings Volume 969), 2008, Vol. 969, pp. 709–716.

Mou, F. L., Xiao, X., Zhang, T., Liu, Q., Li, D. M., Hu, C. W., & Ma, W. (2018). A HIL simulation facility for task verification of the Chinese space station manipulator. Proceedings of 2018 IEEE International Conference on Mechatronics and Automation, 5–8 August, 2018, Changchun, China.

Osaki, K., Konno, A., & Uchiyama, M. (2010). Time delay compensation for a hybrid simulator. Advanced Robotics, 24(8–9), 1081–1098.

Piedboeuf, J., De Carufel, J., Aghili, F., & Dupuis, E. (1999). Task verification facility for the Canadian special purpose dexterous manipulator. Proceedings of the 1999 IEEE International Conference on Robotics and Automation, 10–15 May, 1999, Detroit, MI, USA.

Qi, C. K., Gao, F., Zhao, X. C., Ren, A. Y., Wang, Q., Sun, Q., Hu, Y., & Qiao, L. (2016). Smith predictor based delay compensation for a hardware-in-the-loop docking simulator. Mechatronics, 36, 63–76.

Qi, C. K., Wang, W. L., Li, D. J., Hu, Y., & Gao, F. (2021). Convergence compensation for space contact semiphysical simulator based on mechanical structural dynamics. Journal of Aerospace Engineering, 34(5), 04021066.

Qi, C. K., Li, D. J., Ma, W., Wei, Q. Q., Zhang, W. M., Wang, W. L., Hu, Y., & Gao, F. (2022). Distributed delay compensation for a hybrid simulation system of space manipulator capture. IEEE/ASME Transactions on Mechatronics, 27(4), 2367–2378.

Qi, C. K., Gao, F., Zhao, X. C., Ren, A. Y., & Wang, Q. (2017). A force compensation approach towards divergence of hardware-in-the-loop contact simulation system for damped elastic contact. IEEE Transactions on Industrial Electronics, 64(4), 2933–2943.

Qi, C. K., Ren, A. Y., Gao, F., Zhao, X. C., Wang, Q., & Sun, Q. (2017). Compensation of velocity divergence caused by dynamic response for hardware-in-the-loop docking simulator. IEEE/ASME Transactions on Mechatronics, 22(1), 422–432.

Qi, C. K., Zhao, X. C., Gao, F., Ren, A. Y., & Hu, Y. (2017). Low-order model based divergence compensation for hardware-in-the-loop space discrete contact. Journal of Intelligent and Robotic Systems, 86(1), 81–93.

Qi, C. K., Gao, F., Zhao, X. C., Wang, Q., & Ren, A. Y. (2018). Hybrid smith predictor and phase lead based divergence compensation for hardware-in-the-loop contact simulation with measurement delay. Acta Astronautica, 147, 175–182.

Qi, C. K., Gao, F., Zhao, X. C., Wang, Q., & Sun, Q. (2018). Distortion compensation for a robotic hardware-in-the-loop contact simulator. IEEE Transactions on Control Systems Technology, 26(4), 1170–1179.

Qi, C. K., Li, D. J., Hu, Y., Gao, F., Wang, W. L., & Ma, W. (2023). Stability analysis and distortion compensation of robot structural dynamics for a hybrid simulator. Journal of Mechanical Design, 145(7): 073302.

Qi, C. K., Li, D. J., Hu, Y., Zheng, Y., Wang, W. J., Shou, X., & Gao, F. (2023). Learning-based distortion compensation for a hybrid simulator of space docking. IEEE Robotics and Automation Letters, 8(6): 3446–3453.

Rems, F., Frei, H., Risse, E.-A., & Burri, M. (2021). 10-year anniversary of the European proximity operations simulator 2.0 - looking back at test campaigns, rendezvous research and facility improvements. Aerospace, 8(9), 235.

Riley, D. R., Jaquet, B. M., Pennington, J. E., & Brissenden, R. F. (1966). Comparison of results of two simulations employing full-size visual-cue for pilot-controlled gemini-agena docking. NASA TN D-3687, pp. 1–35.

Romano, M., Friedman, D. A., & Shay, T. J. (2007). Laboratory experimentation of autonomous spacecraft approach and docking to a collaborative target. Journal of Spacecraft and Rockets, 44(1), 164–173.

Sawada, H., Szuki, S., & Oda, M. (2007). A contact dynamics analysis of spacecraft capturing and rigidizing. The 17th Workshop on JAXA Astrodynamics and Flight Mechanics, 2007.

Sazawal, P., Choukroun, D., Benninghoff, H., & Eberhard, G. (2019). Three-dimensional modeling and time-delay stability analysis for robotics docking simulation. Proceedings of the Institution of Mechanical Engineers, Part G: Journal of Aerospace Engineering, 233(14), 5438–5455.

Schwartz, J. L., Peck, M. A., & Hall, C. D. (2003). Historical review of air-bearing spacecraft simulators. Journal of Guidance, Control, and Dynamics, 26(4), 513–522.

Shimoji, H., Inoue, M., Tsuchiya, K., Niomiya, K., Nakatani, I., & Kawaguchi, J. (1991). Simulation system for a space robot using six-axis servos. Advanced Robotics, 6(2), 179–196.

Takahashi, R., Ise, H., Konno, A., Uchiyama, M., & Sato, D. (2008). Hybrid simulation of a dual-arm space robot colliding with a floating object. Proceedings of the 2008 IEEE International Conference on Robotics and Automation, 19–23 May, 2008, Pasadena, CA, USA.

Uyama, N., Fujii, Y., Nagaoka, K., & Yoshida, K. (2012). Experimental evaluation of contact-impact dynamics between a space robot with a compliant wrist and a free-flying object. International Symposium on Artificial Intelligence, Robotics and Automation in Space, 2012, Turin, Italy.

Wang, Q., Qi, C. K., Gao, F., Zhao, X. C., Ren, A. Y., & Sun, Q. (2019). Force based delay compensation for hardware-in-the-loop simulation divergence of 6-dof space contact. Proceedings of the Institution of Mechanical Engineers, Part G: Journal of Aerospace Engineering, 233(1), 151–165.

Xu, W. F., Liang, B., Xu, Y. S., Li, C., & Qiang, W. Y. (2007). A ground experiment system of free-floating space robot for capturing space target. Journal of Intelligent & Robotic Systems, 48(2), 187–208.

Xu, W. F, Liang, B., & Xu, Y. S. (2011). Survey of modeling, planning, and ground verification of space robotic systems. Acta Astronautica, 68, 1629–1649.

Yaskevich, A. (2014). Real time math simulation of contact interaction during spacecraft docking and berthing. Journal of Mechanics Engineering and Automation, 4, 1–15.

Yu, S. M., Han, J. W., Zhang, W. M., & Xu, D. M. (2022). Smith predictor compensation and fuzzy incremental control for delay of space docking hardware-in-the-loop simulation system. Proceedings of the Institution of Mechanical Engineers, Part G: Journal of Aerospace Engineering, 236(11), 2179–2193.

Yu, S. M., Han, J. W., Qu, Z. Y., & Yang, Y. (2018). A force and displacement compensation method toward divergence and accuracy of hardware-in-the-loop simulation system for manipulator docking. IEEE Access, 6, 35091–35104.

Yu, S. M., Han, J. W., Yang, Y., Xu, D. M., & Qu, Z. Y. (2018). Force and moment compensation method based on three degree-of-freedom stiffness-damping identification for manipulator docking hardware-in-the-loop simulation system. IEEE Access, 6, 63452–63467.

Zebenay, M., Boge, T., Krenn, R., & Choukroun, D. (2015). Analytical and experimental stability investigation of a hardware-in-the-loop satellite docking simulator. Proceedings of the Institution of Mechanical Engineers, Part G: Journal of Aerospace Engineering, 229(4), 666–681.

Zebenay, M., Boge, T., & Choukroun, D. (2013). Modeling, stability analysis, and testing of a hybrid docking simulator. AIAA Guidance, Navigation, and Control (GNC) Conference, 19–22 Aug, 2013, Boston, MA, USA.

Zhu, W.-H., Piedboeuf, J.-C., & Gonthier, Y. (2002). Emulation of a space robot using a hydraulic manipulator on ground. Proceedings of the 2002 IEEE International Conference on Robotics and Automation, 11–15 May, 2002, Washington, DC, USA.

Chapter 2
Robot System for HIL Simulation of Space Contact Dynamics

2.1 HIL Simulation Principle

In the space 0-g environment, the active spacecraft carrying the active docking mechanism approaches the passive docking mechanism on the target spacecraft. The active and passive docking mechanisms on the spacecraft are contact hardwares that refer to the physical model in this study. When two spacecraft dock through the docking mechanisms, the contact force will be produced. The contact force will change the motion of two spacecraft according to the dynamics law. Then the new contact force will be produced and two spacecraft will move to the new position. This dynamic process will continue until the completion of the space task. The docking system in space is shown in Fig. 2.1.

On the ground, the physical model is installed on a motion simulator. The motion simulator is used to implement the relative motion of two spacecraft. The force sensor is used to measure the actual contact force. By utilizing the measured contact force as the input, the numerical model (i.e., the dynamics model of the spacecraft) is used to calculate the relative position of two spacecraft (i.e., the desired position of the motion simulator). The motion simulator is controlled to track the desired position. Depending on the actual position of the motion simulator, the physical model will produce the actual contact force. When the simulation task completes or the relative velocity becomes zero, the iterative simulation process will stop. The HIL simulation principle is shown in Fig. 2.2.

The goal of the HIL simulation on the ground is to reproduce the contact dynamics process of two spacecraft with the physical model in the space 0-g environment. The physical model used in the HIL simulation on the ground is the same as that of spacecraft in space. The numerical model used in the HIL simulation on the ground is the dynamics model which describes the relative motion of the spacecraft with respect to the contact force. The space 0-g environment is simulated in a numerical and economic way. The complicated and expensive hardware for the emulation of the space 0-g environment is not required. However, the HIL simulation has much

C. Qi et al., *Control of Hardware-In-The-Loop Space Contact Simulation Robot*,
https://doi.org/10.1007/978-981-95-1172-3_2

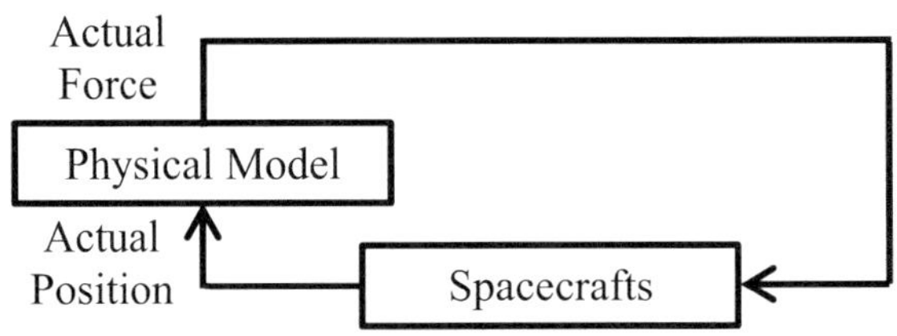

Fig. 2.1 Docking system in space

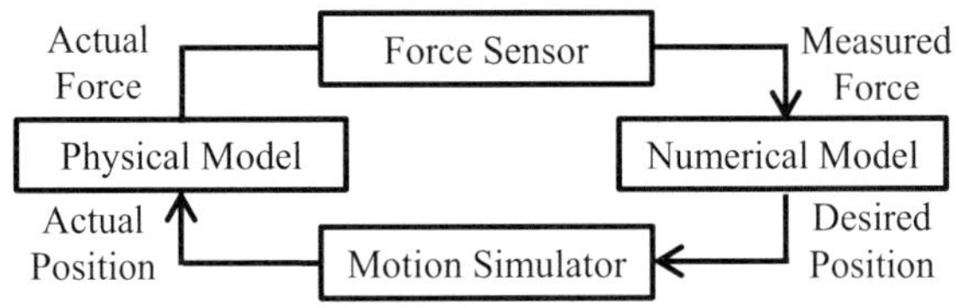

Fig. 2.2 HIL simulation principle

more challenges. It is required to achieve a dynamics similarity of two systems with different structure. The whole dynamics of the motion simulator, physical model, force sensor and numerical model in the HIL simulation system should be similar to the dynamics of the physical model and spacecraft in space.

2.2 HIL Simulation System

According to the HIL simulation principle, a HIL simulation system is established. Figure 2.3 shows the HIL simulation system structure. The system consists of an upper platform, a force sensor, active and passive docking mechanisms (i.e., physical model), a lower motion simulator and a numerical model. The force sensor is installed between the upper platform and the passive docking mechanism. The active docking mechanism is installed on the lower motion simulator. For theoretical analysis and numerical simulation studies, the models of the HIL simulation system will be provided individually in each chapter. Moreover, the models or their mathematical expressions may vary slightly depending on the modeling requirements of each chapter.

Figure 2.4 shows the HIL simulation system. The lower motion simulator is designed by using the novel 3–3 PSS (prismatic-spherical-spherical) orthogonal parallel mechanism (Cao et al. 2015). The novel 3–3 orthogonal parallel mechanism guarantees the performances of the lower motion simulator with the small inertia, high stiffness and high precision. The parallel mechanisms combined with the dual-motor actuators (Zhang et al. 2012) and real-time control systems can realize the high frequency response, high velocity and acceleration of the lower motion simulator. The upper platform is designed by using the 3-PRS (prismatic-revolute-spherical) parallel mechanism. In the simulation of the docking process, the upper platform is fixed at a certain position, while in the simulation of the rendezvous and docking

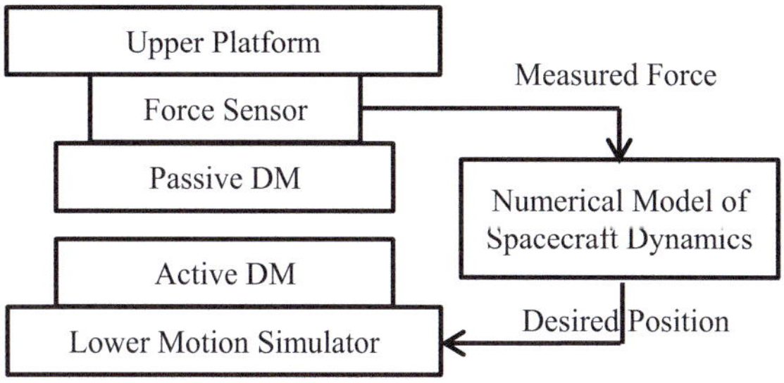

Fig. 2.3 HIL simulation system structure

process, the upper platform is movable in the rendezvous stage and fixed in the docking stage.

The HIL simulation control system is shown in Fig. 2.5. There are three real-time computers which communicate through the fiber-optic real-time network. The numerical simulation computer is used to collect the contact force signals, compute the motion dynamics model of the spacecraft, obtain the desired position of the spacecraft and send it to the motion control computer. The motion control computer is used to communicate with the motor driver through the EtherCAT real-time bus and perform the motion control of the motion simulator. The closed-loop HIL simulation only includes the numerical simulation and motion control systems. The simulation time step is set as 1 ms, which is fast enough for our HIL contact simulation. The monitoring computer is used to monitor the data and states of the HIL contact simulation system. There are two non-real-time computers. The operation control computer is used as the human–machine-interface (HMI) to set the experimental

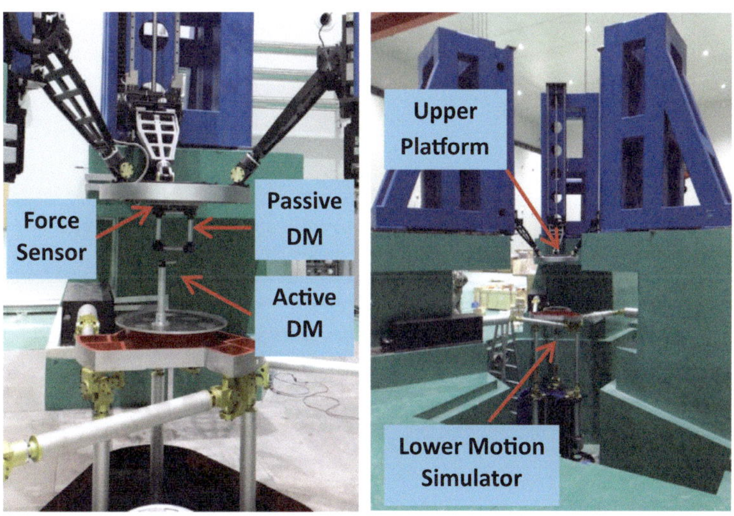

Fig. 2.4 HIL simulation system

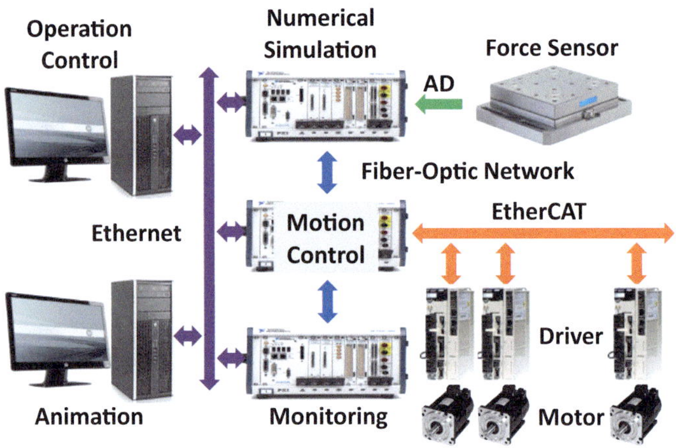

Fig. 2.5 HIL simulation control system

parameters and control the experimental procedure. The animation computer is used to demonstrate the motion animation of the spacecraft and the motion simulator.

2.3 Summary

In this chapter, the principle of the HIL simulation is introduced, and the system structure, mechanical system and control system of the HIL simulation system are presented.

References

Cao, R., Gao, F., Zhang, Y., & Pan, D. (2015). A key point dimensional design method of a 6-DOF parallel manipulator for a given workspace. Mechanism and Machine Theory, 85, 1–13.

Zhang, J. Z., Gao, F., Yu, H. N., & Zhao, X. C. (2012). Use of an orthogonal parallel robot with redundant actuation as an earthquake simulator and its experiments. Proceedings of the Institution of Mechanical Engineers, Part C: Journal of Mechanical Engineering Science, 226(1), 257–272.

Chapter 3
Compensation of Force Measurement Delay

3.1 Introduction

For the space docking HIL simulator, because the bandwidths of the force/torque sensor, signal acquisition and noise filtering are limited, the force measurement system always has a force measurement delay (FMD). It is known that the FMD leads to the velocity divergence of the HIL simulator. The coefficient of restitution (CoR) of velocity is larger than what occurs in space. To improve the simulation accuracy and guarantee the system stability, the FMD must be compensated. Some compensation control studies have considered the FMD (Osaki et al. 2010; Qi et al. 2016; Chang et al. 2007; Qi et al. 2017a, c; Wang et al. 2019).

When the time delay value is known, a first-order phase lead compensator can be obtained by setting the phase lead time to be the delay time (Shimoji et al. 1991; Osaki et al. 2010; Qi et al. 2016; Qi et al. 2017a). When the delay model and contact stiffness are known, a first-order compensator can be designed based on the model to achieve the desired CoR (Osaki et al. 2010; Abiko et al. 2014). When the delay model is unknown, a first-order compensator can be designed using the trial-and-error method (Chang et al. 2007; Zebenay et al. 2015).

To compensate the time delay, Smith predictor is widely used in the process control field (Zhong and Li 2002; Wang et al. 2000). However, in the HIL contact simulation field, few studies are reported on the use of the Smith predictor. In Qi et al. (2016), it is found that for the pure time delay, a first-order phase lead compensator still results in a divergence. To overcome this problem, a serial Smith predictor and phase lead compensation approach is proposed (Qi et al. 2016), which only requires the time delay value.

In fact, the serial Smith predictor and phase lead compensation in Qi et al. (2016) still leads to a small convergence, which becomes more obvious when the contact natural frequency increases. The simulation accuracy could be further improved, which is the motivation of this study. To achieve a higher simulation accuracy with a little convergence, a hybrid Smith predictor and phase lead compensation is proposed

C. Qi et al., *Control of Hardware-In-The-Loop Space Contact Simulation Robot*,
https://doi.org/10.1007/978-981-95-1172-3_3

(Qi et al. 2018). It only requires the time delay value, and does not require the contact model of the docking mechanisms. The phase angle of the compensator is analyzed and the stability condition of the closed-loop system is given. The performance of the proposed divergence compensation approach is tested by simulations on an elastic contact process.

The rest of the chapter is organized as follows. The modeling of the HIL simulation system is presented in Sect. 3.2. The divergence phenomena of the HIL simulation are given in Sect. 3.3. The force measurement delay compensation approaches are presented in Sect. 3.4. The phase analyses and stability conditions are given in Sect. 3.5. Finally, a summary is provided in Sect. 3.6.

3.2 Models of HIL Simulation System

For the simulation studies, the mathematical models of the HIL simulation system are given as shown in Fig. 3.1.

3.2.1 Contact Model

The contact model is used to describe the relationship between the contact force and the relative position. Here, we only consider the one-dimensional contact for simplicity though the proposed method can also be extended to multi-dimensional contact. Moreover, only the undamped elastic contact is considered. It means that the contact is similar to a mass-spring system, where there is always a rebound velocity after the contact. In fact, the proposed method can also be applicable to the damped elastic contact. Finally, the continuous contact is considered because the discrete contacts can be converted to multiple continuous contacts.

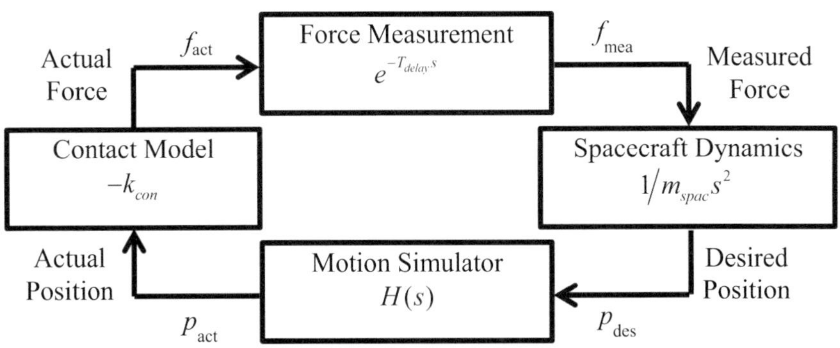

Fig. 3.1 HIL simulation system models

Under the above assumptions, the contact model can be expressed as

$$f_{act}(t) = -k_{con}p_{act}(t) \tag{3.1}$$

where $f_{act}(t)$ is the actual contact force, k_{con} is the contact stiffness, $p_{act}(t)$ is the actual relative position of docking mechanisms. Of course, the contact damping can also be considered in the model. However, the results are similar and the conclusions are not changed. Therefore, for simplicity, here the contact model only considers the contact stiffness.

3.2.2 Force Measurement

The force measurement system includes the force sensor, amplifier, data acquisition card and noise filter. Considering all of them, there is a time delay between the measured force and actual contact force. The force measurement delay can be estimated from the sensor delay, sampling time and filter delay. It can also be identified from the time difference between the measured force and the actual position because the actual position has the same phase as the actual contact force. The force measurement delay is assumed to a pure time delay as follows

$$F_{mea}(s) = e^{-T_{\text{delay}}s}F_{act}(s) \tag{3.2}$$

where $F_{mea}(s)$ is the Laplace transform of the actual force $f_{mea}(t)$, $F_{act}(s)$ is the Laplace transform of the actual force $f_{act}(t)$, and T_{delay} is the force measurement delay. In some cases, the pure time delay of the dynamic response of the motion simulator can be combined with the force measurement delay together for the analysis.

3.2.3 Spacecraft Dynamics Model

As a numerical model, the spacecraft dynamics model is used to calculate the numerical relative motion of spacecraft from the contact force as follows

$$m_{spac}d^2p_{num}(t)/dt^2 = f_{num}(t) \tag{3.3}$$

where m_{spac} denotes the equivalent mass of two spacecraft in space, $p_{num}(t)$ is the numerical relative position of two spacecraft, $f_{num}(t)$ is the contact force used for the numerical computation of the dynamics model. Note that $f_{num}(t)$ is equal to $f_{mea}(t)$ when the force measurement delay compensation is not used. The relative position $p_{num}(t)$ is sent to the motion simulator as the desired position $p_{des}(t)$. For the contact process, the contact natural frequency can be derived as $f_{con} = 1/2\pi\sqrt{k_{con}/m_{spac}}$.

3.2.4 Motion Simulator

The motion simulator is used to reproduce the relative motion of spacecraft. Due to the limited response capability, there is always a dynamic response delay of the motion simulator, i.e., a dynamic delay between the actually arrived position $p_{act}(t)$ and the desired position $p_{des}(t)$. The dynamic response delay can be compensated by the model-based feedforward position compensation (Qi et al. 2017c) or the response error based force compensation (Qi et al. 2017b). When the dynamic response delay can be approximated by a pure time delay T_{resp} as follows

$$p_{act}(t) = p_{des}(t - T_{resp}) \tag{3.4}$$

this time delay can be combined with the force measurement delay for the analysis. It means that the force measurement delay T_{delay} in Eq. 3.2 also includes the dynamic response delay T_{resp} of the motion simulator. Therefore, in the following analysis we can only focus on the force measurement delay compensation, and the response model of the motion simulator is given as follows

$$P_{act}(s) = H(s)P_{des}(s) = P_{des}(s) \tag{3.5}$$

where $H(s) = 1$.

3.3 Divergence Phenomenon of HIL Simulation

In the following, we will show that the HIL simulation is divergent due to the time delay. As an example, the simulation settings are given as follows: the equivalent mass 500kg, the initial velocity 100mm/s, and the time delay 7ms. Three contact frequencies 2Hz, 5Hz and 8Hz are selected, and the corresponding contact stiffness is 78.96, 493.48 and 1263.31N/mm respectively. These contact parameters are selected because they are close to real space contacts considered in this study. Of course, other reasonable parameters can also be selected, however similar results will be achieved and the conclusions do not change.

In an ideal case, the CoR is 1 for the undamped elastic contact. Figure 3.2 shows the desired velocity without the force measurement delay compensation and Fig. 3.3 shows the corresponding CoR. It can be seen that the HIL simulation without the force measurement delay compensation has a larger divergence as the contact natural frequency increases. The corresponding CoR at 2Hz, 5Hz and 8Hz contact natural frequency is 1.1478, 1.4060 and 1.7067 respectively. For the HIL contact simulator, the system divergence is not permitted because it could destroy the tested docking mechanisms. Therefore, the force measurement delay compensation is very necessary.

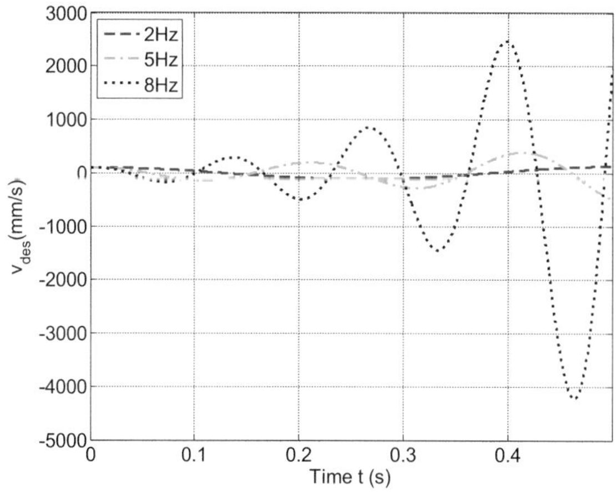

Fig. 3.2 Desired velocity without force measurement delay compensation

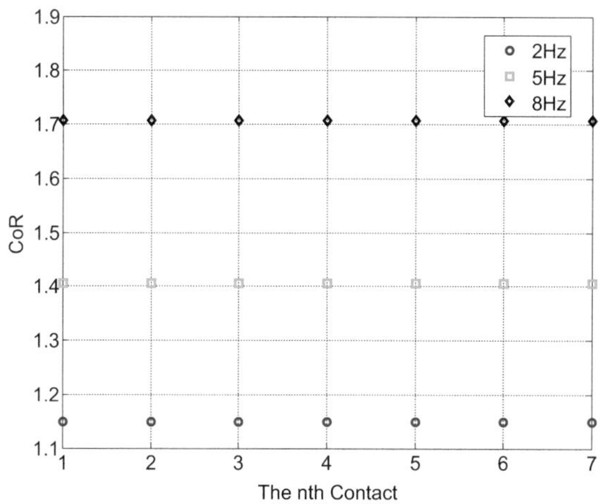

Fig. 3.3 CoR without force measurement delay compensation

3.4 Force Measurement Delay Compensation Approach

3.4.1 Phase Lead Compensation

As shown in the Eq. 3.2, we assume that the force measurement delay is a pure delay, i.e., $e^{-T_{\text{delay}}s}$. In theory, the time lead $e^{T_{\text{delay}}s}$ is a perfect compensation. In the implementation, the phase lead compensation is performed by using a first-order Taylor expansion of $e^{T_{\text{delay}}s}$

$$F_{\text{lead}}(s) = (1 + T_{\text{lead}}s)F_{mea}(s) \tag{3.6}$$

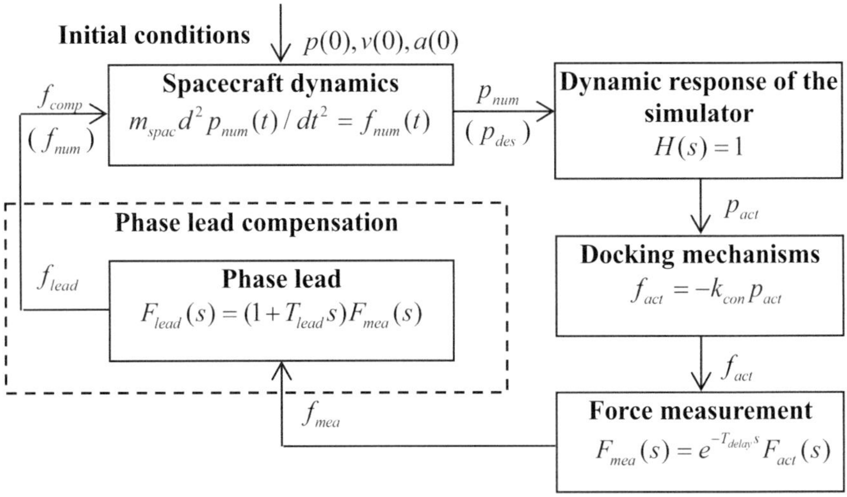

Fig. 3.4 Phase lead compensation

where $F_{\text{lead}}(s)$ is the Laplace transform of the force $f_{\text{lead}}(t)$ after phase lead compensation, and T_{lead} is the time lead. The time lead T_{lead} is often set to be equal to the delay value T_{delay}. The phase lead compensation approach is shown in Fig. 3.4.

To test the phase lead compensator, the HIL simulation is performed, where the equivalent mass is 500kg, the initial velocity is 100mm/s and the time delay is 7ms, and the time lead is 7ms.

Figure 3.5 shows the desired velocity at the contact natural frequency 2Hz, 5Hz and 8Hz, and the corresponding CoR is shown in Fig. 3.6. For the 2Hz contact, the divergence compensation is acceptable since the CoR is decreased from 1.1478 to 1.0004. For the 5Hz contact, there is a small divergence, where the CoR is 1.0058. For the 8Hz contact, the simulation divergence becomes obvious, where the CoR is 1.0250. The first-order phase lead cannot compensate the simulation divergence very well. The simulation becomes more divergent when the contact natural frequency increases.

3.4.2 Serial Smith Predictor and Phase Lead Compensation

As shown in Fig. 3.7, the ideal Smith predictor compensation is

$$\Delta F_{\text{smith}}(s) = (1 - e^{-T_{\text{smith}}s})F_{act}(s) \tag{3.7}$$

In the implementation, a first-order Pade approximation of $e^{-T_{\text{smith}}s}$ is often used

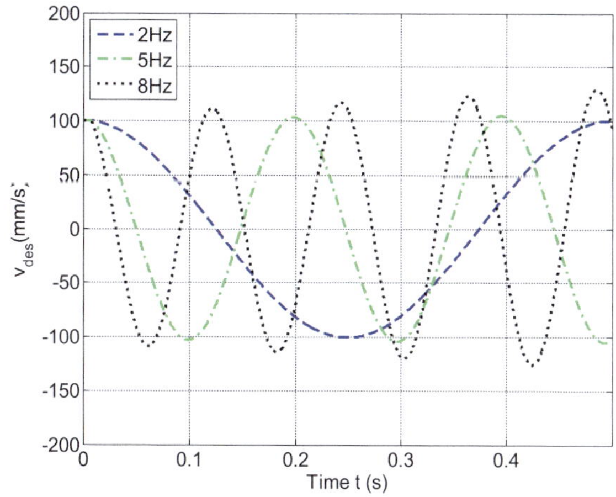

Fig. 3.5 Desired velocity with phase lead compensation

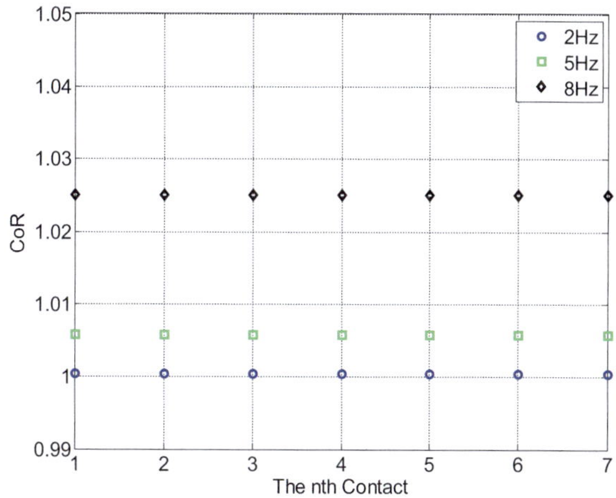

Fig. 3.6 CoR with phase lead compensation

$$e^{-T_{\text{smith}}s} = \frac{1 - T_{\text{smith}}s/2}{1 + T_{\text{smith}}s/2} \tag{3.8}$$

The input force f_{act} of the Smith predictor is unknown, and it is approximated by a predicted value f_{lead} of the phase lead in Eq. 3.6. Therefore, the serial Smith predictor and phase lead compensation is given by Qi et al. (2016)

$$F_{\text{smith}}(s) = (1 + \frac{T_{\text{smith}}s}{1 + T_{\text{smith}}s/2}(1 + T_{\text{lead}}s))F_{mea}(s) \tag{3.9}$$

The serial Smith predictor and phase lead compensation is shown in Fig. 3.8. It integrates a Smith predictor and a phase lead, where the output of the phase lead is the input of the Smith predictor.

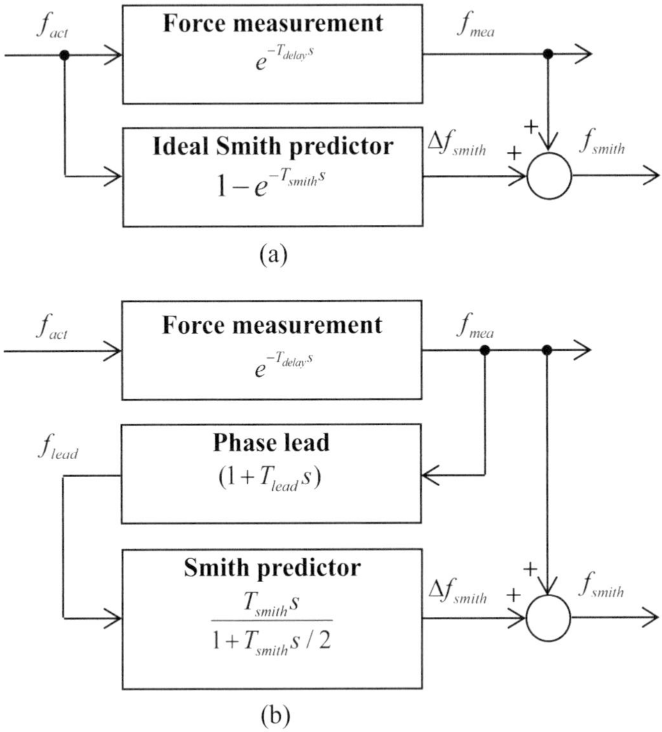

Fig. 3.7 Smith predictor compensation principle

The simulation are performed to test the serial Smith predictor and phase lead compensation, where the spacecraft mass is 500kg, the contact velocity is 100mm/s and the time delay is 7ms. The phase lead and Smith predictor are $T_{\text{lead}} = 7\text{ms}$ and $T_{\text{smith}} = 7\text{ms}$.

Figure 3.9 shows the desired velocity at the 2, 5, 8Hz contact natural frequency and Fig. 3.10 shows the corresponding CoR. The results can be compared with those of the phase lead compensation in Figs. 3.5 and 3.6. For the 2Hz contact, the serial Smith predictor and phase lead compensator works very well, where the CoR is 0.9996. For the 5Hz contact, the simulation is convergent because the CoRs is 0.9933. For the 8Hz contact, the simulation has an obvious convergence, and the CoR is 0.9743. It can been seen that the serial Smith predictor and phase lead compensation approach will lead to an over-compensation when the contact natural frequency increases.

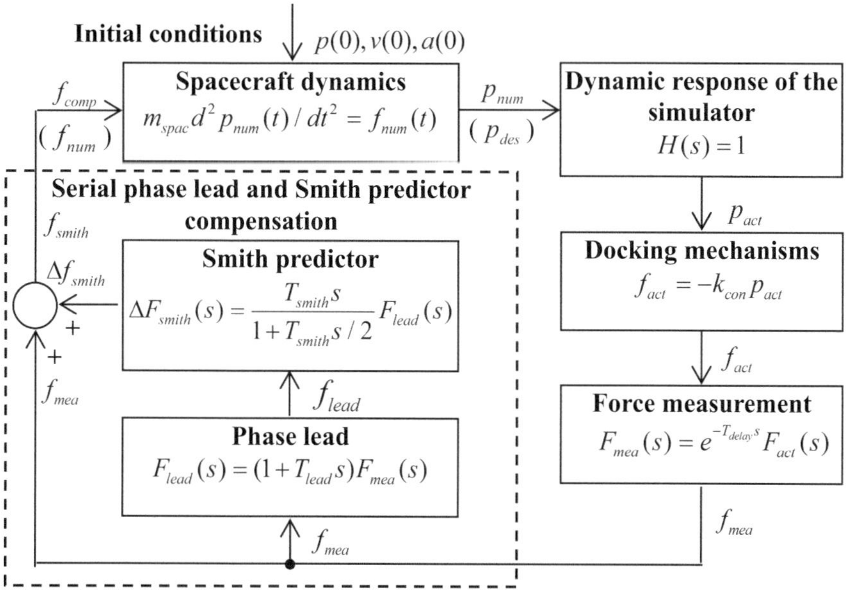

Fig. 3.8 Serial Smith predictor and phase lead compensation

Fig. 3.9 Desired velocity with serial Smith predictor and phase lead compensation

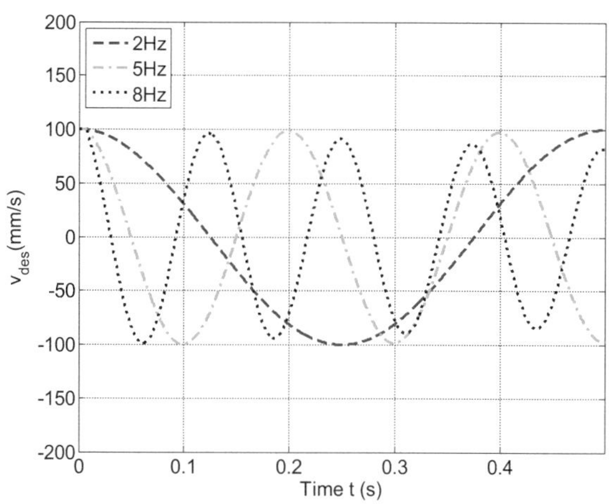

3.4.3 *Hybrid Smith Predictor and Phase Lead Compensation*

From the above studies, it can be seen that the phase lead compensation results in the simulation divergence and the serial phase lead and Smith predictor compensation results in the simulation convergence. The divergence and convergence are obvious when the contact natural frequency increases. If we can integrate these two compensation approaches, the simulation could achieve the balance between the divergence

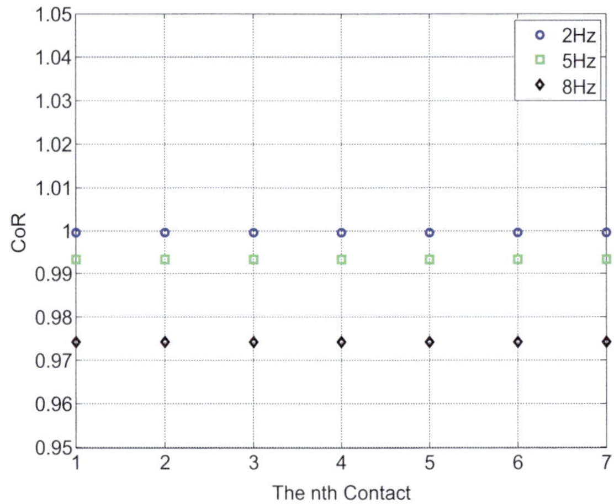

Fig. 3.10 CoR with serial Smith predictor and phase lead compensation

and the convergence. As shown in Fig. 3.11, the proposed hybrid Smith predictor and phase lead compensation takes average on the phase lead compensation and the serial Smith predictor and phase lead compensation

$$f_{\text{hybrid}} = (f_{\text{lead}} + f_{\text{smith}})/2 \tag{3.10}$$

That is

$$F_{\text{hybrid}}(s) = \left(\frac{1 + T_{\text{lead}}s}{2} + \frac{1 + 3T_{\text{smith}}s/2 + T_{\text{smith}}T_{\text{lead}}s^2}{2 + T_{\text{smith}}s}\right)F_{mea}(s) \tag{3.11}$$

The hybrid Smith predictor and phase lead compensation is tested by the simulation, where the spacecraft mass is 500kg, the contact velocity is 100mm/s and the time delay is 7ms. The phase lead and Smith predictor are $T_{\text{lead}} = 7\text{ms}$ and $T_{\text{smith}} = 7\text{ms}$. These simulation conditions are the same as the previous simulations in order to compare three force measurement delay compensation approaches clearly.

Figure 3.12 gives the desired velocity and Fig. 3.13 shows the corresponding CoR. The CoR of three compensation approaches are compared in Table 3.1. Obviously, for the 2Hz contact, three compensation approaches all work very well because the CoR is reduced from 1.1478 to 1.0004, 0.9996 and 1.0000 respectively. For the 5Hz contact, with the hybrid Smith predictor and phase lead, the simulation convergence is very small because the CoR is 0.9994. However, the CoR with the phase lead compensation and with the serial Smith predictor and phase lead compensation are 1.0058 and 0.9933 respectively. For the 8Hz contact, with the hybrid Smith predictor and phase lead compensation, the simulation convergence is also very satisfactory because the CoR is 0.9985. But the CoR with the phase lead compensation and with the serial Smith predictor and phase lead compensation are 1.0250, 0.9743 respectively.

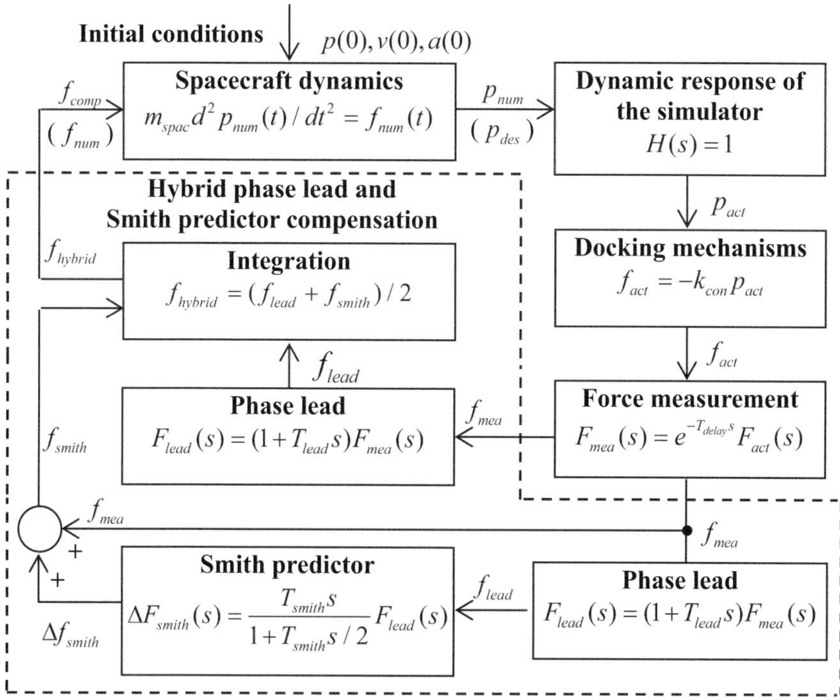

Fig. 3.11 Hybrid Smith predictor and phase lead compensation

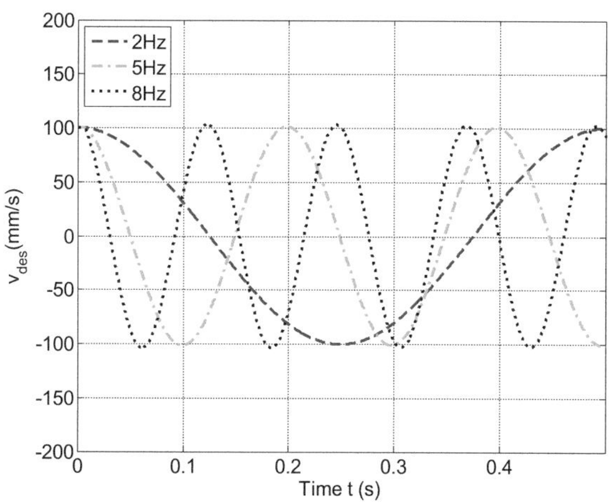

Fig. 3.12 Desired velocity with hybrid Smith predictor and phase lead compensation

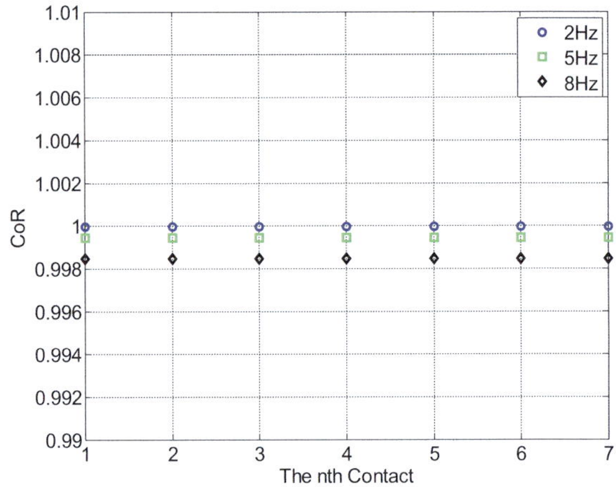

Fig. 3.13 CoR with hybrid Smith predictor and phase lead compensation

Table 3.1 CoR with three compensation approaches

Contact natural frequency (Hz)	Without compensation	Phase lead compensation	Serial Smith predictor and phase lead compensation	Hybrid Smith predictor and phase lead compensation	CoR ideal value
2	1.1478	1.0004	0.9996	1.0000	1
5	1.4060	1.0058	0.9933	0.9994	1
8	1.7067	1.0250	0.9743	0.9985	1

The simulations on different equivalent mass, different initial velocity and different time delay can also achieve similar results, which are ignored due to limited space. Now we can summarize that as the contact natural frequency increases, the phase lead compensation will lead to an obvious divergence, and the serial Smith predictor and phase lead compensation will result in an obvious convergence. Fortunately, the proposed hybrid Smith predictor and phase lead compensation can achieve very satisfactory results. The proposed method is effective for the force measurement delay compensation of the HIL contact simulation.

3.5 Phase and Stability Analysis of Compensation Approaches

In order to explain why the compensation approaches can reduce the divergence, the phase angle of the compensator is analyzed as follows.

The phase angle of the ideal compensator is $T_{\text{delay}}\omega$, where $\omega = 2\pi f_{con}$ is the contact angle frequency.

For the first-order phase lead compensator, the phase angle between the compensated force and the measured force is

$$\angle\frac{F_{comp}(s)}{F_{mea}(s)} = \tan^{-1}(T_{\text{lead}}\omega) \tag{3.12}$$

Figure 3.14 shows the phase angle of the ideal compensator and the phase lead compensator at 4ms, 7ms and 10ms time delay and at 1-11Hz contact natural frequency. It can be understood that when the contact natural frequency or the time delay increases, the compensated phase angle also increases. The first-order phase lead compensator still leads to a simulation divergence, because its phase angle is less than that of the ideal compensator. To guarantee the stable and convergent simulation, the phase angle of the compensator should be a little bit larger than that of the ideal compensator.

For the serial Smith predictor and phase lead compensation, the transfer function between the compensated force and the measured force is

$$\frac{F_{comp}(s)}{F_{mea}(s)} = 1 + \frac{T_{\text{smith}}s}{1 + T_{\text{smith}}s/2}(1 + T_{\text{lead}}s) \tag{3.13}$$

and the phase angle is

$$\angle\frac{F_{comp}(s)}{F_{mea}(s)} = \tan^{-1}\left(\frac{3T_{\text{smith}}\omega/2}{1 - T_{\text{smith}}T_{\text{lead}}\omega^2}\right) - \tan^{-1}(T_{\text{smith}}\omega/2) \tag{3.14}$$

Figure 3.14 also shows the phase angle of the serial Smith predictor and phase lead compensation. Because its phase angle is larger than that of the ideal compensation, the simulation is convergent.

For the hybrid Smith predictor and phase lead compensation, the transfer function between the compensated force and the measured force is

$$\frac{F_{comp}(s)}{F_{mea}(s)} = \frac{1 + T_{\text{lead}}s}{2} + \frac{1 + 3T_{\text{smith}}s/2 + T_{\text{smith}}T_{\text{lead}}s^2}{2 + T_{\text{smith}}s} \tag{3.15}$$

and the phase angle is

$$\angle\frac{F_{comp}(s)}{F_{mea}(s)} = \tan^{-1}\left(\frac{(T_{\text{smith}} + T_{\text{lead}}/2)\omega}{1 - 3T_{\text{smith}}T_{\text{lead}}\omega^2/4}\right) - \tan^{-1}(T_{\text{smith}}\omega/2) \tag{3.16}$$

Figure 3.14 also shows the phase angle of the hybrid Smith predictor and phase lead compensation. It can be seen that its phase angle is between those of the serial Smith predictor and phase lead compensation and the phase lead compensation. The simulation is convergent, when its phase angle is smaller than that of the ideal compensation, for example at several milliseconds time delay and several Hertz contact natural frequency. These convergence regions cover interested contact cases

Fig. 3.14 Compensation phase analysis

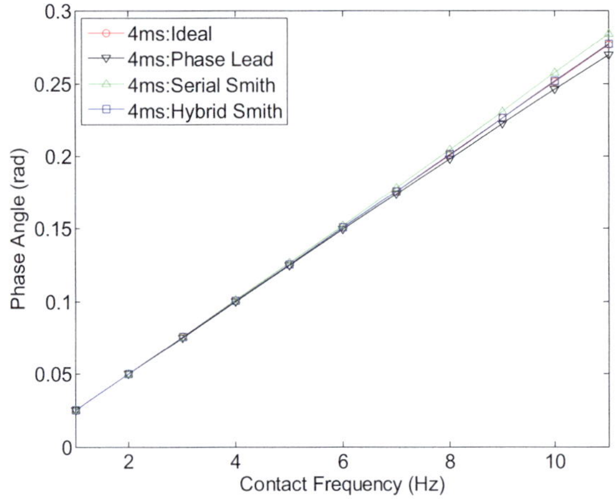

(a) Phase angle at 4ms time delay

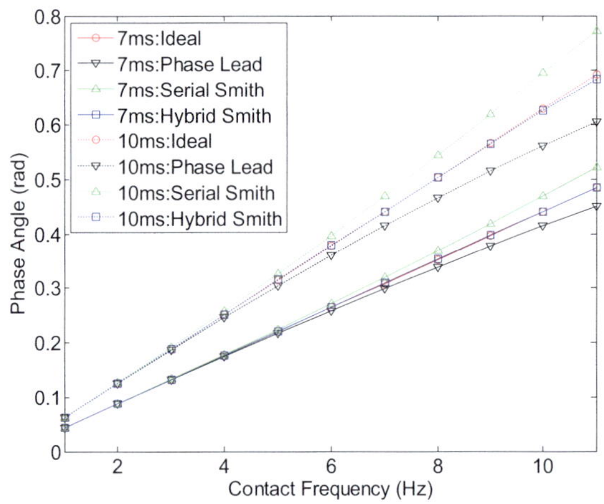

(b) Phase angle at 7ms and 10ms time delay

considered in this study. Though the hybrid Smith predictor and phase lead compensation can achieve a higher fidelity, the simulation could be divergent as the time delay or the contact natural frequency increase. Therefore, stability conditions on the time delay and the contact natural frequency are necessary.

The stability condition of the hybrid Smith predictor and phase lead compensation can be analyzed as follows. For simplicity, we assume that the Smith predictor and phase lead have the same time parameters as the time delay: $T_{\text{smith}} = T_{\text{lead}} = T_{\text{delay}}$. For the closed-loop HIL simulation system, its characteristic equation can be derived from the Eqs 3.1–3.5 and 3.11 as follows

$$\chi(s) = m_{spac}s^2(1 + \frac{1}{2}T_{\text{delay}}s) + e^{-T_{delay}s}k_{con}(1 + \frac{3}{2}T_{\text{delay}}s + \frac{3}{4}T_{\text{delay}}^2 s^2) \qquad (3.17)$$

In the stability analysis, a first-order Pade approximation as shown in Eq. 3.8 is used to rationalize the pure time delay. After that, the characteristic equation is

$$\chi(s) = \frac{1}{2} m_{spac} T_{\text{delay}}^2 s^4 + (2m_{spac} T_{\text{delay}} - \frac{3}{4} k_{con} T_{\text{delay}}^3) s^3 \\ +2m_{spac} s^2 + 2k_{con} T_{\text{delay}} s + 2k_{con} \tag{3.18}$$

The closed-loop stability conditions (necessary and sufficient conditions) can be derived using Routh-Hurwitz rules as follows

$$T_{\text{delay}} < \sqrt{\frac{8}{9} \frac{m_{spac}}{k_{con}}} = \frac{\sqrt{2}}{3} \frac{1}{\pi f_{con}} \tag{3.19}$$

Obviously, the contact natural frequency and the time delay are limited by the condition 3.19 in order to maintain the stability. If the condition 3.19 is satisfied, the closed-loop HIL contact simulation system is stable.

When the HIL simulator has some nonlinear dynamics, the time delay compensation and stability analysis becomes more complicated. The delay compensation and analysis for the linear HIL system can be extended when the nonlinear HIL system is linearized at one working point. Another way is to use the control and stability analysis methods for nonlinear systems with time delay. It is worth investigating in the future.

3.6 Summary

The hardware-in-the-loop simulation is suitable to test the contact dynamics of the docking process of two spacecraft on the ground. The HIL simulation has a divergent problem due to the time delay. For the pure time delay of the force measurement system, the first-order phase lead compensation still results in a divergence, while the serial Smith predictor and phase lead compensation achieves a small convergence. The divergence or convergence becomes more obvious as the contact natural frequency increases. In this chapter, the proposed hybrid Smith predictor and phase lead compensation can achieve a little convergence and a higher simulation fidelity. The phase analysis and stability condition are given. The performances of the proposed compensation approaches are tested and compared by the simulations on the elastic contact process.

References

Abiko, S., Satake, Y., Jiang, X., Tsujita, T., & Uchiyama, M. (2014). Time delay compensation based on coefficient of restitution for collision hybrid motion simulator. Advanced Robotics, 28(17), 1177–1188.

Chang, T., Cong, D., Ye, Z., & Han, J. (2007). Time problems in HIL simulation for on-orbit docking and compensation. 2nd IEEE Conference on Industrial Electronics and Applications, 23–25 May, 2007, Harbin, China.

Osaki, K., Konno, A., & Uchiyama, M. (2010). Time delay compensation for a hybrid simulator. Advanced Robotics, 24(8–9), 1081–1098.

Qi, C. K., Gao, F., Zhao, X. C., Ren, A. Y., Wang, Q., Sun, Q., Hu, Y., & Qiao, L. (2016). Smith predictor based delay compensation for a hardware-in-the-loop docking simulator. Mechatronics, 36, 63–76.

Qi, C. K., Gao, F., Zhao, X. C., Wang, Q., & Ren, A. Y. (2018). Hybrid smith predictor and phase lead based divergence compensation for hardware-in-the-loop contact simulation with measurement delay. Acta Astronautica, 147, 175–182.

Qi, C. K., Gao, F., Zhao, X. C., Ren, A. Y., & Wang, Q. (2017). A force compensation approach towards divergence of hardware-in-the-loop contact simulation system for damped elastic contact. IEEE Transactions on Industrial Electronics, 64(4), 2933–2943.

Qi, C. K., Ren, A. Y., Gao, F., Zhao, X. C., Wang, Q., & Sun, Q. (2017). Compensation of velocity divergence caused by dynamic response for hardware-in-the-loop docking simulator. IEEE/ASME Transactions on Mechatronics, 22(1), 422–432.

Qi, C. K., Zhao, X. C., Gao, F., Ren, A. Y., & Hu, Y. (2017). Low-order model based divergence compensation for hardware-in-the-loop space discrete contact. Journal of Intelligent & Robotic Systems, 86(1), 81–93.

Shimoji, H., Inoue, M., Tsuchiya, K., Niomiya, K., Nakatani, I., & Kawaguchi, J. (1991). Simulation system for a space robot using six-axis servos. Advanced Robotics, 6(2), 179–196.

Wang, Q., Qi, C. K., Gao, F., Zhao, X. C., Ren, A. Y., & Sun, Q. (2019). Force based delay compensation for hardware-in-the-loop simulation divergence of 6-dof space contact. Proceedings of the Institution of Mechanical Engineers, Part G: Journal of Aerospace Engineering, 233(1), 151–165.

Wang, Q. G., Bi, Q., & Zhang, Y. (2000). Re-design of Smith predictor systems for performance enhancement. ISA Transactions, 39, 79–92.

Zebenay, M., Boge, T., Krenn, R., & Choukroun, D. (2015). Analytical and experimental stability investigation of a hardware-in-the-loop satellite docking simulator. Proceedings of the Institution of Mechanical Engineers, Part G: Journal of Aerospace Engineering, 229(4), 666–681.

Zhong, Q. C., & Li, H.-X. (2002). 2-degree-of-freedom proportional-integral- derivative-type controller incorporating the Smith principle for processes with dead time. Industrial Engineering and Chemistry Research, 41, 2448–2454.

Chapter 4
Compensation of Dynamic Response Delay of Lower Motion Simulator

4.1 Introduction

For the space docking HIL simulator, due to limited bandwidth of the actuation system and external inertia, the lower motion simulator generates the dynamic response delay (DRD) during tracking the motion command. It means that the actual position of the lower motion simulator is delayed to the desired position of the lower motion simulator (i.e., the numerical position computed by the dynamics model of the spacecraft). Then the phase of the actual contact force is delayed to the desired motion of the spacecraft. As a result, the DRD of the lower motion simulator leads to the velocity divergence of the HIL simulator. To obtain a high fidelity of simulation, the DRD of the lower motion simulator should be compensated. Some compensation control studies have considered the DRD of the lower motion simulator (Ananthakrishnan et al. 1996; Qi et al. 2017a, b).

The model-based feed-forward position compensation (Ananthakrishnan et al. 1996) can be used to correct the command position of motion simulator and make its actual position approximate the desired position computed by the spacecraft dynamics model. The position command compensator can be designed using a low-order model (Qi et al. 2017b). The above delay compensation approaches assume that the DRD value of the lower motion simulator or delay model is known, however in some cases it is difficult to obtain the accurate the delay model due to the complicated time-varying dynamics of the docking simulator.

In this chapter, we consider the dynamic response delay of the motion simulator. Because the dynamic response of the motion simulator is related to the posture of the motion simulator, the contact force and the contact natural frequency, it is difficult to model its complicated time-varying dynamics. This motivates the model-free based delay compensation for the HIL docking simulator. A response error real-time feedback based force compensation approach is proposed to handle the dynamic response delay of the motion simulator (Qi et al. 2017a). The compensation force is obtained from the response error of the motion simulator and the contact

C. Qi et al., *Control of Hardware-In-The-Loop Space Contact Simulation Robot*,
https://doi.org/10.1007/978-981-95-1172-3_4

stiffness and damping. The contact stiffness and damping are identified in real-time from the contact force, the actual position and velocity. Because this compensation approach does not require the dynamic response model of the motion simulator, it can be easily applied in the HIL docking simulator. Simulations and experiments are used to verify the effectiveness of the proposed compensation approach.

The rest of this study is organized as follows. The HIL simulation modeling is presented in Sect. 4.2. The HIL simulation divergence is discussed in Sect. 4.3. The HIL simulation compensation is given in Sect. 4.4. Section 4.5 reports simulations and experiments. Finally, a summary is presented in Sect. 4.6.

4.2 HIL Simulation Modeling

For the analysis, the mathematical model of the HIL simulation system is established as shown in Fig. 4.1. In the following section, each model will be given.

4.2.1 Physical Model

For simplicity, we assume that the active spacecraft perpendicularly collides with the target spacecraft, and that the interaction between two spacecraft can be expressed as a mass-spring-damper system without the friction. Under these assumptions, the actual contact force $f_{act}(t)$ acting on the active DM can be expressed as

$$f_{act}(t) = -k_{con}p_{act}(t) - b_{con}\dot{p}_{act}(t) \tag{4.1}$$

where k_{con} is the contact stiffness, b_{con} is the contact damping, $p_{act}(t)$ and $\dot{p}_{act}(t)$ is the actual relative position and velocity of the docking mechanisms at time t, $p_{act} = 0$ when two spacecraft contact at the spring natural length.

Fig. 4.1 HIL simulation modeling

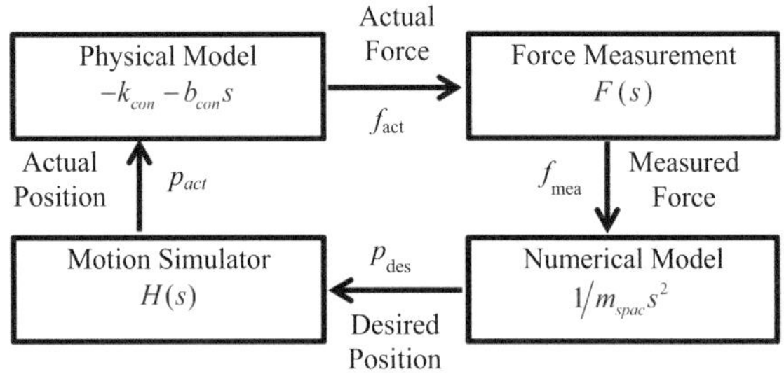

4.2.2 Force Measurement

Considering the hardware of the force sensor and amplifier, the buffer of the data acquisition card and the filtering of the force signals, there is a time delay between the measured force $f_{mea}(t)$ and the actual force $f_{act}(t)$. Here assume that the force measurement system can be described as a low-pass filter

$$F_{mea}(s) = F(s)F_{act}(s) = \frac{1}{1+Ts}F_{act}(s) \tag{4.2}$$

In this study, T = 0.002 s and the bandwidth is 79.6Hz. It is enough because the bandwidth is much larger than the several Hz contact natural frequency. Because this study is focused on the delay compensation caused by the dynamic response of the motion simulator, we assume that the time delay of the force measurement system has been compensated very well. To do this, the force measurement delay (FMD) compensator is designed as the inverse of the low-pass filter

$$W(s) = 1 + Ts \tag{4.3}$$

Of course, other force measurement delay models such as pure time delay can also be used. Because the high-performance compensators for the force measurement delay can be designed. Now, for simplicity, the force measurement after the delay compensation is assumed to be ideal without the time delay. In this case, we have $W(s)F(s) = 1$, therefore

$$\tilde{F}_{mea}(s) = W(s)F(s)F_{act}(s) = F_{act}(s) \tag{4.4}$$

4.2.3 Numerical Model

The measured force is then input to the numerical model in the computer. The numerical relative position $p_{num}(t)$ of two spacecraft for the next time is calculated from the Newton's second law

$$m_{spac}d^2p_{num}(t)/dt^2 = f_{num}(t) \tag{4.5}$$

where m_{spac} denotes the equivalent mass of two spacecraft in space, $f_{num}(t)$ is the contact force used for the numerical computation of the dynamics model, and $f_{num}(t)$ is the same as $f_{mea}(t)$ when there is no delay compensation and $f_{comp}(t)$ when there is the delay compensation. The numerical position $p_{num}(t)$ computed from the numerical model is sent to the motion simulator as the desired position $p_{des}(t)$ of the motion simulator. From the contact stiffness and the equivalent mass of two spacecraft in space, the contact natural frequency can be derived $f_{con} = 1/2\pi\sqrt{k_{con}/m_{spac}}$.

4.2.4 Motion Simulator

The motion simulator demonstrates the actual position $p_{act}(t)$. Due to the limited bandwidth of the motion simulator, there is the tracking error between the actual position and the desired position of the motion simulator. We assume the dynamic response between the actual position and the desired position of the motion simulator can be described by a second-order model

$$p_{act}(t) = H(s)p_{des}(t) = \frac{\omega_{resp}^2}{s^2 + 2\xi_{resp}\omega_{resp}s + \omega_{resp}^2}p_{des}(t) \tag{4.6}$$

where ω_{resp} is the undamped natural frequency (rad/s), ξ_{resp} is the damping ratio. The response frequency with the unit of Hz is denoted as $f_{resp} = \omega_{resp}/2\pi$. Note that this second-order model is only used to describe the dynamic response of the motion simulator in the simulation study, while the proposed HIL compensation method does not require this model.

4.3 HIL Simulation Divergence

In the following section, we show the HIL simulation divergence phenomenon due to the dynamic response delay of the motion simulator. This response delay is due to the limited dynamic response capability of the motion simulator (see the Eq. 4.6). The rebound velocity is larger than the approach velocity, i.e., CoR > 1. The HIL simulation is more divergent when the frequency response of the motion simulator is lower. The HIL simulation divergence mechanism is also discussed.

4.3.1 HIL Simulation Divergence Phenomenon

Now we study the effect of the dynamic response delay of the motion simulator on the HIL simulation. In the simulation, the contact stiffness is 100N/mm, the contact damping is 0.1N/(mm/s) and the initial velocity is 100mm/s. The sampling time and control period are 1ms.

Firstly, the simulation on the 2Hz contact natural frequency is performed. Figure 4.2 shows the ideal velocity and the desired velocity without the compensation at the 100, 50 and 25Hz frequency response of the motion simulator. Figure 4.3 shows the CoR without the compensation at the ideal, 100, 50 and 25Hz frequency response of the motion simulator. It can be seen that at the 2Hz contact natural frequency the ideal simulation is convergent (CoR = 0.9805) and the actual simulation without the compensation is divergent. Moreover, the simulation divergence becomes larger when the frequency response of the motion simulator is lower because the CoR is 1.0310, 1.0841 and 1.1976 respectively.

Secondly, the simulation on the 5Hz contact natural frequency is performed. Figure 4.4 presents the ideal velocity and the desired velocity without the compensation at the 100, 50 and 25Hz frequency response of the motion simulator. As shown in Fig. 4.5, the CoR without the compensation at the ideal, 100, 50 and 25Hz frequency response of the motion simulator is 0.9518, 1.0793, 1.2222 and 1.5499 respectively. Obviously, the ideal simulation is convergent due to the damping, while the actual

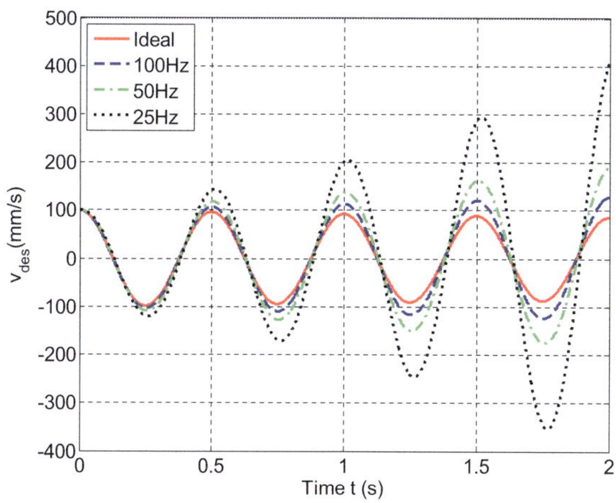

Fig. 4.2 Ideal velocity and desired velocity without compensation at 2Hz contact natural frequency

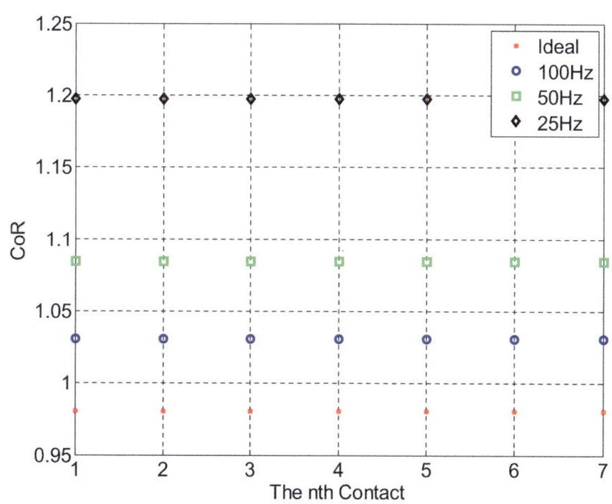

Fig. 4.3 CoR without compensation at 2Hz contact natural frequency

simulation has the significant divergence. Meanwhile, the simulation is more divergent when the frequency response of the motion simulator is reduced. Compared with the 2Hz contact natural frequency, the simulation divergence at the 5Hz contact natural frequency is larger.

Figures 4.2, 4.3, 4.4 and 4.5 show that: (a) the ideal damped elastic contact in space without the time delay is convergent. (b) the actual HIL simulation for the damped elastic contact with the time delay is divergent. This means that the HIL divergence is due to the time delay.

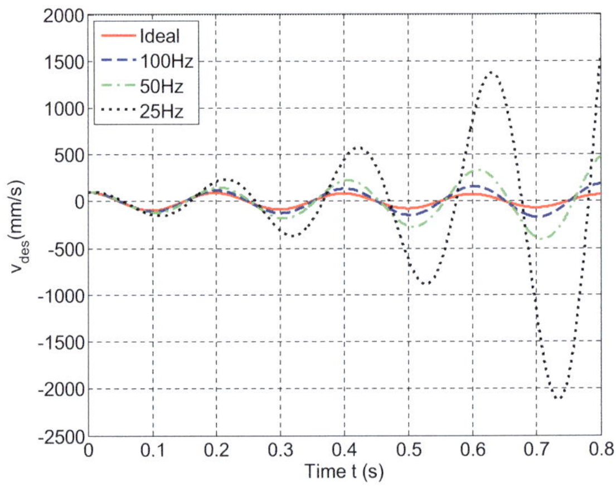

Fig. 4.4 Ideal velocity and desired velocity without compensation at 5Hz contact natural frequency

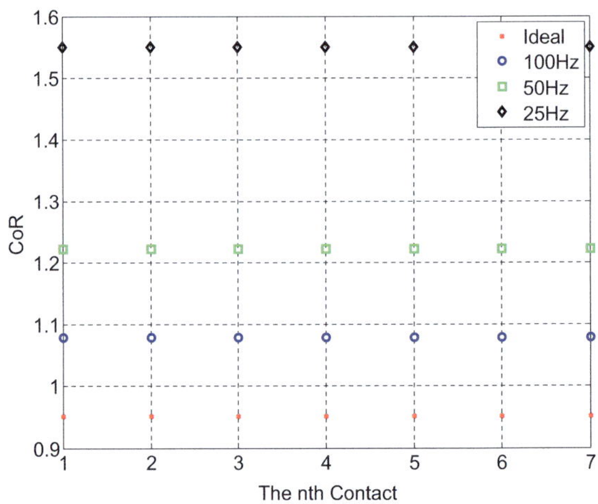

Fig. 4.5 CoR without compensation at 5Hz contact natural frequency

4.3.2 HIL Simulation Divergence Mechanism

The HIL simulation divergence mechanism can be discovered from the desired position, actual position, measured force and desired velocity as shown in Fig. 4.6. It can be seen that the actual position is delayed with respective to the desired position due to the dynamic response of the motion simulator. Before the desired position reaches to the peak value, the desired velocity gradually decreases to zero due to the resistance force, which is called the striking and decelerating process. After the desired position returns from the peak value, the desired velocity gradually increases due to the rebound force, which is called the rebounding and accelerating process. The work done by the resistance force during the striking and decelerating process of the spacecraft in space is S_{in}, and the work done by the rebound force during the rebounding and accelerating process is S_{out}. When $S_{out} > S_{in}$, the simulation divergence happens. This is the HIL simulation divergence mechanism.

4.4 Dynamic Response Delay Compensation Approach

In the following, we propose the response error feedback based force compensation and discuss the identification of contact stiffness and damping. The HIL simulation compensation process is given.

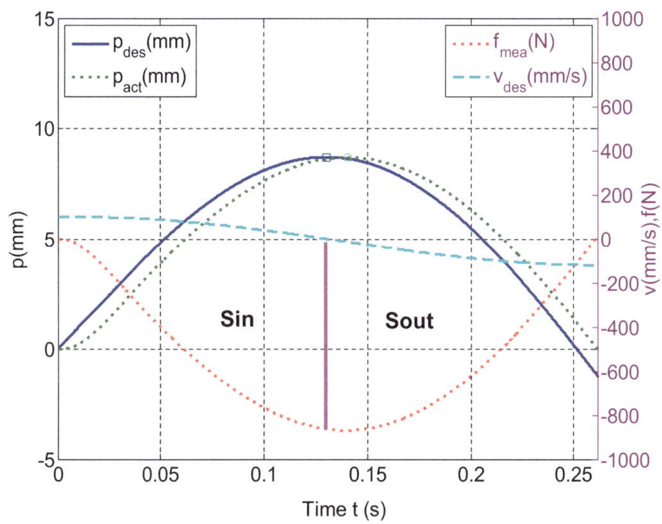

Fig. 4.6 HIL simulation divergence mechanism

4.4.1 Response Error Feedback Based Force Compensation

The response error between the desired position $p_{des}(t)$ and the actual position $p_{act}(t)$, the desired velocity $\dot{p}_{des}(t)$ and the actual velocity $\dot{p}_{act}(t)$ will lead to the contact force error. The force error will further result in the desired position and velocity error. Because the HIL simulation is a closed-loop simulation, the simulation error will accumulate and the system will diverge. This motivates to compensate the contact force to avoid the simulation divergence. Because the dynamic response of the motion simulator is related to the posture of the motion simulator, the contact force and the contact natural frequency, it is difficult to model its complicated time-varying dynamics. Therefore, the model-free based dynamic response compensation is necessary.

In this study, a response error real-time feedback based force compensation approach is proposed as shown in Fig. 4.7. The dynamic response model of the motion simulator is not required.

The position and velocity response error of the simulator is

$$\Delta p_{resp}(t) = p_{des}(t) - p_{act}(t) \tag{4.7}$$

Similarly, the velocity response error is given by

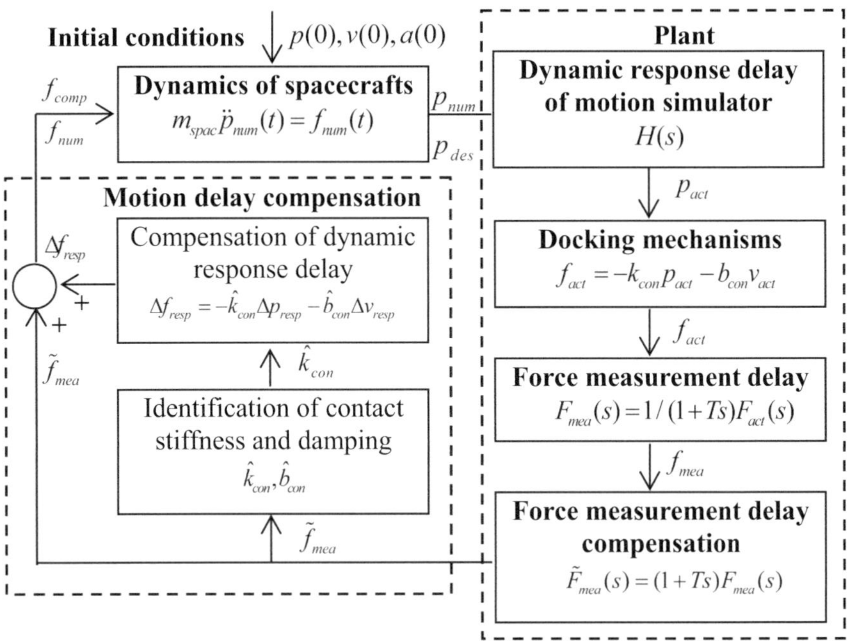

Fig. 4.7 Response error feedback based force compensation

$$\Delta v_{resp}(t) = dp_{des}(t)/dt - dp_{act}(t)/dt \tag{4.8}$$

This position response error and velocity response error will lead to the contact force error $\Delta f_{resp}(t)$.

The physical characteristics including the stiffness and damping of the contact mechanism determine the relationship between the actual position, actual velocity and contact force

$$f_{act}(t) = z_{con}(p_{act}(t)) \tag{4.9}$$

In this study, it is assumed that in a very short period (e.g., several milliseconds) the physical characteristic of the contact mechanism does not change. By real-time identifying and utilizing the physical characteristic of the contact mechanism, it is possible to compute the compensation force. Motivated by the Taylor expansion, the force increment corresponding to the position and velocity increment can be written as

$$\Delta f_{pact}(t) = \left.\frac{\partial z_{con}}{\partial p_{act}}\right|_{p_{act}(t)} \Delta p_{act}(t) = -k_{con}(t)\Delta p_{act}(t) \tag{4.10}$$

$$\Delta f_{vact}(t) = \left.\frac{\partial z_{con}}{\partial v_{act}}\right|_{v_{act}(t)} \Delta v_{act}(t) = -b_{con}(t)\Delta v_{act}(t) \tag{4.11}$$

Thus, the compensation force corresponding to the position response error and velocity response error can be expressed by

$$\Delta f_{resp}(t) = -\hat{k}_{con}(t)\Delta p_{resp}(t) - \hat{b}_{con}(t)\Delta v_{resp}(t) \tag{4.12}$$

where $\hat{k}_{con}(t)$ is the estimated contact stiffness, and $\hat{b}_{con}(t)$ is the estimated contact damping.

4.4.2 Identification of the Contact Stiffness and Damping

In the simulation, the physical characteristic of the contact mechanism is unknown. In this study, we assume the contact mechanism has only the stiffness and damping without the friction for simplicity. The problem is to identify unknown contact stiffness and damping in real-time.

From the Eq. 4.10, we have

$$\Delta f_{act}(t) = -k_{con}(t)\Delta p_{act}(t) - b_{con}(t)\Delta v_{act}(t) \tag{4.13}$$

where $\Delta f_{act}(t)$, $\Delta p_{act}(t)$ and $\Delta v_{act}(t)$ are incremental values of the contact force, actual position and actual velocity. When a set of actual position $p_{act}(t)$,$p_{act}(t -$

1),$\cdots$, $p_{act}(t-n)$, actual velocity $v_{act}(t)$,$v_{act}(t-1)$,...,$v_{act}(t-n)$ and the actual contact force $f_{act}(t)$,$f_{act}(t-1)$,$\cdots$,$f_{act}(t-n)$ are available, where n is the number of sampled data points, $\Delta f_{act}(t)$, $\Delta p_{act}(t)$ and $\Delta v_{act}(t)$ can be constructed.

$$\Delta p_i^t = -\Delta p_{act}(t-n+i) = p_{act}(t-n+i-1) - p_{act}(t-n+i) \tag{4.14}$$

$$\Delta v_i^t = -\Delta v_{act}(t-n+i) = v_{act}(t-n+i-1) - v_{act}(t-n+i) \tag{4.15}$$

$$\Delta f_i^t = \Delta f_{act}(t-n+i) = f_{act}(t-n+i) - f_{act}(t-n+i-1) \tag{4.16}$$

It is assumed that at the n time periods the contact stiffness and damping does not change. In our experiment, n = 2 or 3 is enough to get the accurate and real-time stiffness and damping. Then the contact stiffness $k_{con}(t)$ and damping $b_{con}(t)$ can be estimated online in real-time from the data $\{\Delta p_i^t, \Delta v_i^t, \Delta f_i^t\}_{i=1}^n$ using the recursive least squares method (Ljung 1999) as follows

$$\theta_i^t = \theta_{i-1}^t + L_i^t(\Delta f_i^t - (\phi_i^t)^T\theta_{i-1}^t), i = 1, ..., n \tag{4.17}$$

$$L_i^t = \frac{R_{i-1}^t\phi_i^t}{1 + (\phi_i^t)^T R_{i-1}^t \phi_i^t} \tag{4.18}$$

$$R_i^t = R_{i-1}^t - \frac{R_{i-1}^t\phi_i^t(\phi_i^t)^T R_{i-1}^t}{1 + (\phi_i^t)^T R_{i-1}^t \phi_i^t} \tag{4.19}$$

where $\theta_i^t = [k_i^t, b_i^t]^T$ is the estimated stiffness and damping at the ith step for the time t, L_i^t is the updating gain, R_i^t denotes the inverse covariance that is a positive number and $\phi_i^t = [\Delta p_i^t, \Delta v_i^t]^T$, $\hat{k}_{con}(t) = k_n^t$, $\hat{b}_{con}(t) = b_n^t$.

4.4.3 HIL Simulation Compensation Process

As shown in Fig. 4.7, the detailed HIL simulation process is given as follows.

(1) Given the initial conditions (i.e., initial position $p(0)$, velocity $v(0)$ and acceleration $a(0)$), the motion simulator plans and realizes the initial conditions. Set the initial conditions to the numerical model and set $t = 0$.
(2) At the time t, the desired position $p_{des}(t)$ calculated by the numerical simulation is sent to the motion simulator. The motion simulator realizes the desired position. The actual arrival position is $p_{act}(t)$.
(3) The contact mechanism generates the actual contact force $f_{act}(t)$ depending on the actual position and velocity, and the stiffness and damping of the contact mechanism.

(4) The measured force $f_{mea}(t)$ is obtained by through the force and torque sensor, the data acquisition system and measurement delay compensator.
(5) The contact stiffness k_{con} and damping b_{con} of the docking mechanisms are estimated by using the actual contact force $f_{act}(t)$(i.e., the measured contact force after the force measurement delay compensation $\tilde{f}_{mea}(t)$ in this case) and the actual position $p_{act}(t)$ and velocity $v_{act}(t)$.
(6) Using the desired position $p_{des}(t)$ calculated by the numerical simulation and the actual position $p_{act}(t)$ of the motion simulator, the position and velocity response error is obtained. Based on the estimated contact stiffness k_{con} and damping b_{con} of the contact mechanism, the compensation force Δf_{resp} is obtained.
(7) By combining the measured force $\tilde{f}_{mea}(t)$ from the force measurement system after the force measurement delay compensation and the compensation force $\Delta f_{resp}(t)$ for the simulator response, the totally compensated contact force $f_{comp}(t)$ is obtained. It is actually the estimation value of the contact force corresponding to the desired position $p_{des}(t)$ calculated from the numerical simulation.
(8) The desired position $p_{des}(t + \Delta t)$ of the motion simulator for the next time is calculated by the numerical model of the dynamics of two spacecraft. Set $t = t + \Delta t$ and repeat steps 2–8. The simulation stops until the completion of the simulation tasks.

4.5 Verifications

4.5.1 Verifications by Software

To verify the performance of the proposed delay compensation approach for the HIL simulation system, the simulations with the damped elastic contact are performed using the Matlab software. In the simulation, the simulation parameters are given as follows: the contact stiffness $k_{con} = 100$N/mm, the contact damping $b_{con} = 0.1$N/(mm/s) and the initial velocity $v_{num}(0) = 100$mm/s. The initial stiffness estimation is set as $\hat{k}_{con} = 80$N/mm and the initial damping estimation is set as $\hat{b}_{con} = 0.12$N/(mm/s). The sampling time and control period are 1ms. The simulations on two contact frequencies are shown as an example. Note that in the simulations and experiments, the contact stiffness and damping, initial velocity and contact natural frequency are selected to be close to the real space contact. For the considered docking process in space, the contact stiffness is from tens to about one hundred N/mm, the initial velocity is from tens to about one hundred mm/s, and the contact natural frequency is about several Hz.

Firstly, the simulation on the 2Hz contact natural frequency is performed. Figure 4.8 shows the ideal velocity and the desired velocity with the dynamic response compensation at the 100, 50 and 25Hz frequency response. Figure 4.9 gives the CoR with the dynamic response compensation at the 100, 50 and 25Hz frequency response. It can be seen that at the 2Hz contact natural frequency the simulation divergence as shown in Figs. 4.2 and 4.3 can be compensated very well by the dynamic response compensation because the CoR is reduced to about 0.9805 for three cases of frequency response. Figure 4.10 presents the desired position, actual position, desired velocity, measured force, ideal force and compensated force with dynamic response compensation at the 25Hz frequency response as an example. To show the time delay more clearly, a small window of figure is zoomed in for the first contact. It can be seen that the compensated force with the dynamic response compensation is very close to the ideal force. That is why it can achieve good compensation performance.

Secondly, the simulation on the 5Hz contact natural frequency is given. Figure 4.11 presents the ideal velocity and the desired velocity with the dynamic response compensation at the 100, 50 and 25Hz frequency response. Figure 4.12 illustrates the CoR with the dynamic response compensation at the 100, 50 and 25Hz frequency response. The results can be compared with Figs. 4.4 and 4.5. It can be seen that at the

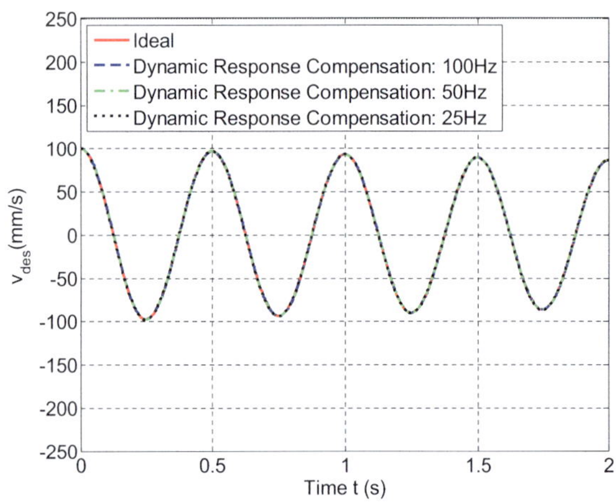

Fig. 4.8 Ideal velocity and desired velocity with dynamic response compensation at 2Hz contact natural frequency

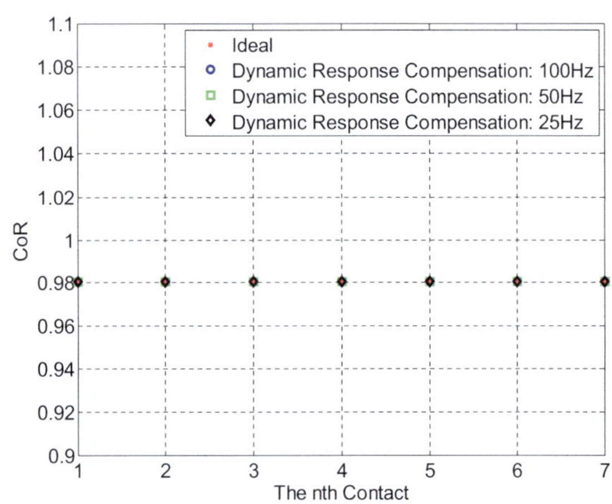

Fig. 4.9 CoR with dynamic response compensation at 2Hz contact natural frequency

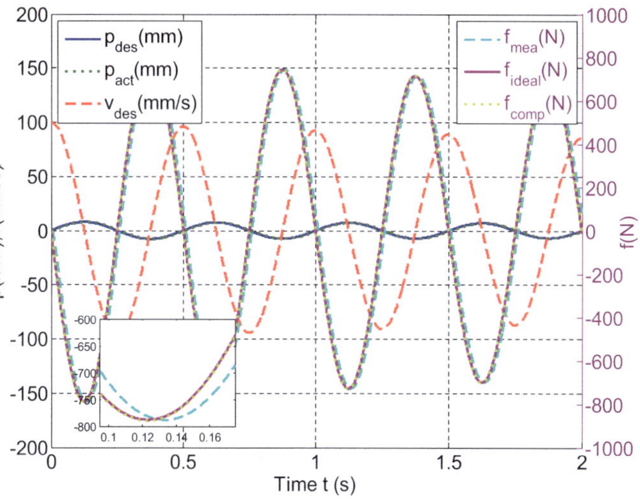

Fig. 4.10 Desired position, actual position, desired velocity, measured force, ideal force and compensated force with dynamic response compensation at 2Hz contact natural frequency and 25Hz frequency response

5Hz contact natural frequency there is also a good simulation performance with the dynamic response compensation because their CoR is about 0.9518. The estimated contact stiffness and damping from initial values tend to the true values as shown in Fig. 4.13. As shown in Fig. 4.14, the compensated force with the dynamic response compensation is very close to the ideal force. Therefore, the HIL simulation with the dynamic response compensation works well.

Fig. 4.11 Ideal velocity and desired velocity with dynamic response compensation at 5Hz contact natural frequency

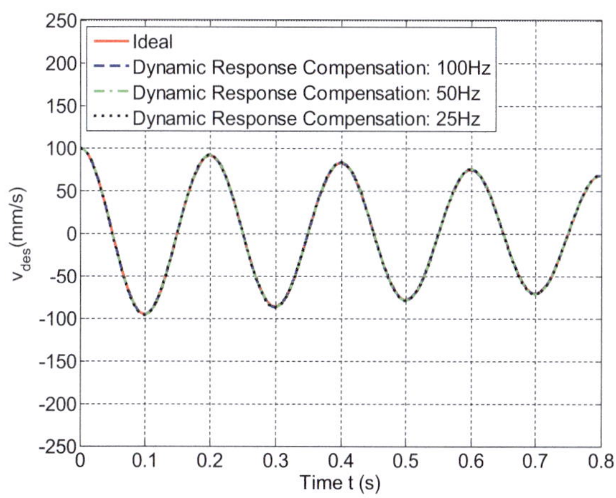

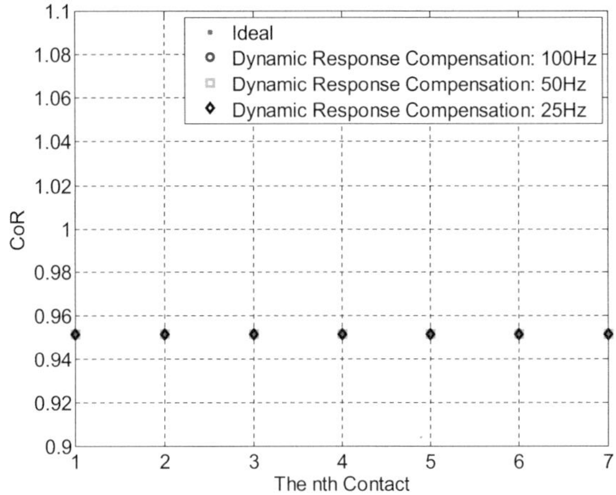

Fig. 4.12 CoR with dynamic response compensation at 5Hz contact natural frequency

4.5.2 Verifications by Experiments

Using the developed HIL simulator, the experiments on a real elastic rod are performed to test the proposed delay compensation approach. Figure 4.15 shows some pictures in the experiments. The four pictures from A to D denote the initial position, the contact with the lower frame, the middle position, the contact with the upper frame, respectively. Because the velocity is uniform between the lower frame and the upper frame, these data are removed and the remaining data are connected in order to show the figures more clearly and compactly. The simulations on two contact frequencies are shown as an example.

For the contact experiments at the 2Hz contact natural frequency, the contact stiffness is $k_{con} = 48.68N/mm$, the equivalent mass is $m_{spac} = 308.27kg$ and the initial velocity is $v(0) = 50mm/s$. Figure 4.16 gives the desired velocity without the compensation and with the dynamic response compensation. As shown in Fig. 4.17, the CoR without the compensation and with the dynamic response compensation is 1.0509 and 0.9816 respectively. It can be seen that without the compensation, the simulation is divergent, and with the dynamic response compensation, the simulation becomes a little convergent. This is expected because it can protect the product safety with the permissible small convergence error. It means that the dynamic response compensation is effective for the HIL simulation at the 2Hz contact natural frequency.

For the contact experiments at the 3Hz contact natural frequency, the contact stiffness is $k_{con} = 103.57N/mm$, the equivalent mass is $m_{spac} = 291.5kg$ and the initial velocity is $v(0) = 50mm/s$. Figure 4.18 presents the desired velocity without the compensation and with the dynamic response compensation. As shown in Fig. 4.19, the CoR is 1.0918 and 0.9809 respectively. Compared with the result at the 2Hz contact natural frequency, the simulation without the compensation becomes more

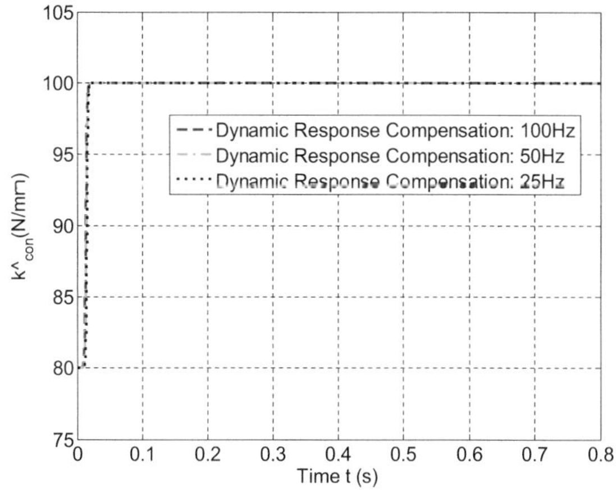

(a)

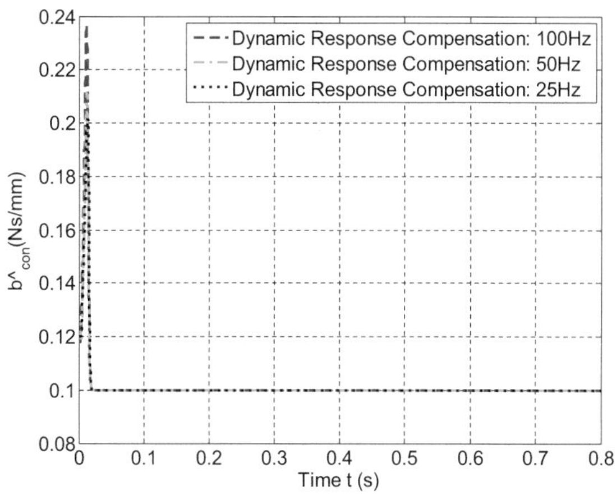

(b)

Fig. 4.13 Estimated contact stiffness and damping with dynamic response compensation at 5Hz contact natural frequency and 25Hz frequency response

divergent. For the dynamic response compensation, the simulation has a small convergence rather than the divergence. The simulation performance is satisfactory because the CoR is also less than and close to one.

The effectiveness of the proposed dynamic response compensation approach for the HIL simulation is verified by both simulations and experiments. The proposed compensation approach can compensate the dynamic delay of the dynamic response of the motion simulator. The HIL simulation divergence problem caused by the dynamic response delay is solved well. Using the proposed compensation approach, the HIL simulation system becomes more accurate and much safer during the testing of the real docking mechanisms.

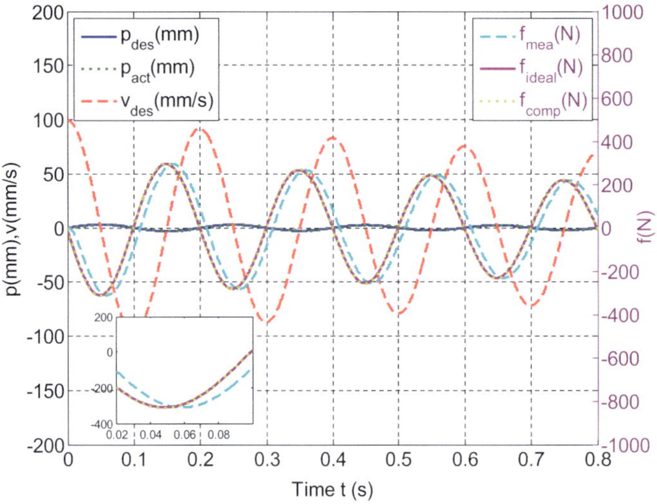

Fig. 4.14 Desired position, actual position, desired velocity, measured force, ideal force and compensated force with dynamic response compensation at 5Hz contact natural frequency and 25Hz frequency response

Fig. 4.15 Contact experiments

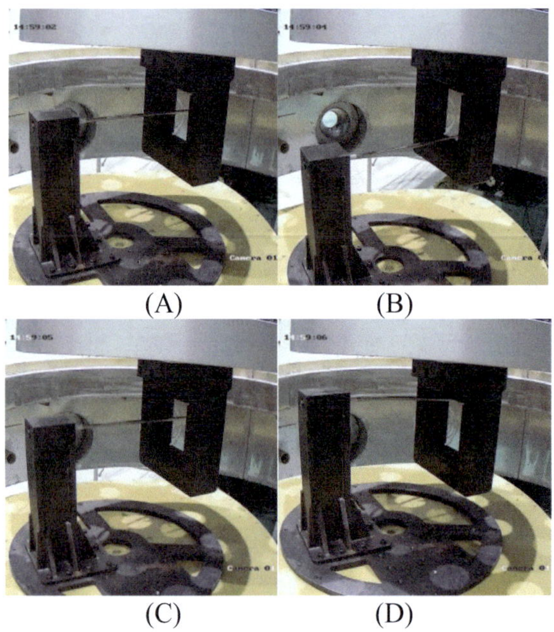

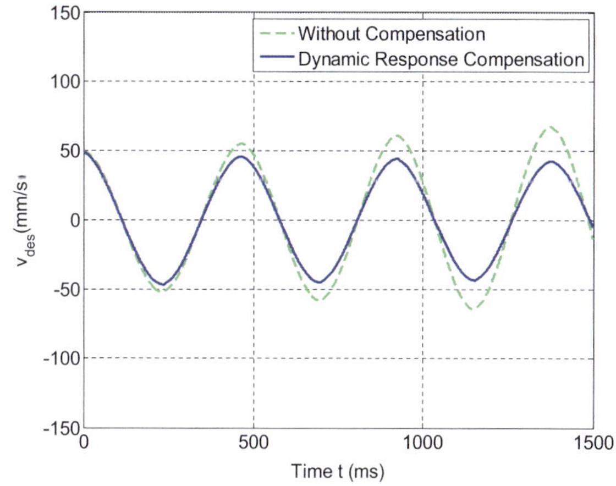

Fig. 4.16 Desired velocity without compensation and with dynamic response compensation at 2Hz contact natural frequency

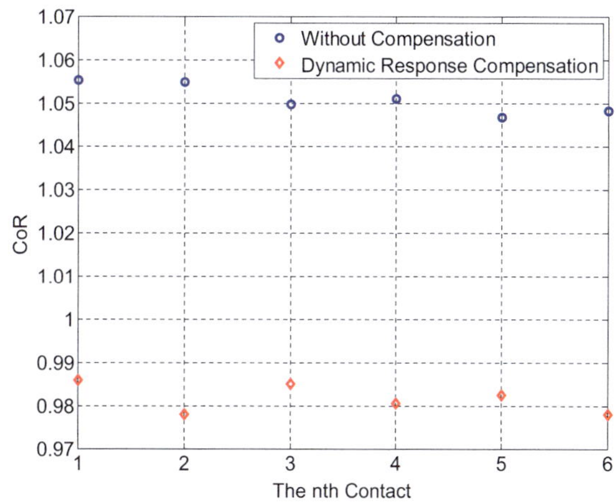

Fig. 4.17 CoR without compensation and with dynamic response compensation at 2Hz contact natural frequency

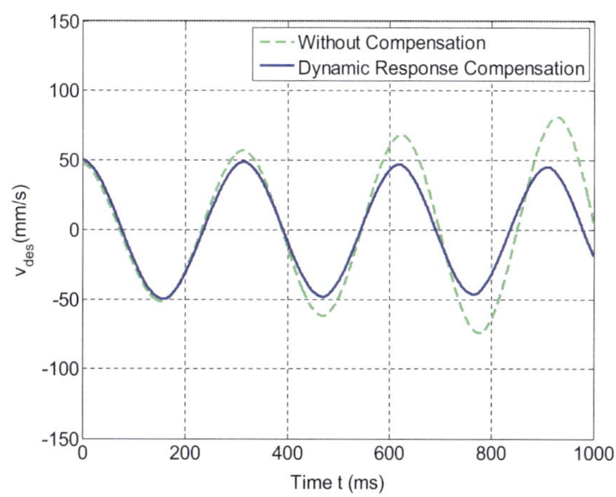

Fig. 4.18 Desired velocity without compensation and with dynamic response compensation at 3Hz contact natural frequency

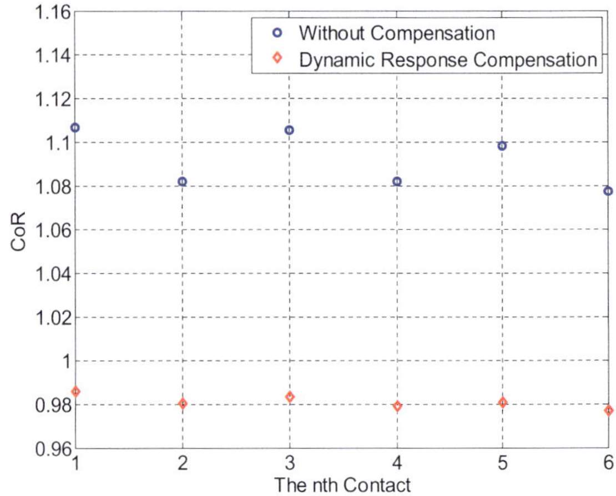

Fig. 4.19 CoR without compensation and with dynamic response compensation at 3Hz contact natural frequency

4.6 Summary

In this chapter, the compensation of the velocity divergence caused by the dynamic response of the motion simulator is investigated. The idea is to compensate the contact force error corresponding to the response error of the motion simulator. The compensation force is obtained from the response error of the motion simulator, the contact stiffness and damping. The contact stiffness and damping are identified in real time from the measured contact force, the actual position and velocity. The proposed compensation approach does not require the dynamic response model of the motion simulator. The compensation approach is practical and useable when the accurate model of the dynamic response is difficult to obtain due to the time-varying and unknown uncertainties. The effectiveness of the proposed compensation approach is verified by the simulations and experiments.

References

Ananthakrishnan, S., Teders, R., & Alder, K. (1996). Role of estimation in real-time contact dynamics enhancement of space station engineering facility. IEEE Robotics and Automation Magazine, 3(3), 20–28.

Ljung, L. (1999). System identification: Theory for the user (the 2nd edition). Englewood Cliffs, NJ: Prentice-Hall.

Qi, C. K., Ren, A. Y., Gao, F., Zhao, X. C., Wang, Q., & Sun, Q. (2017). Compensation of velocity divergence caused by dynamic response for hardware-in-the-loop docking simulator. IEEE/ASME Transactions on Mechatronics, 22(1), 422–432.

Qi, C. K., Zhao, X. C., Gao, F., Ren, A. Y., & Hu, Y. (2017). Low-order model based divergence compensation for hardware-in-the-loop space discrete contact. Journal of Intelligent and Robotic Systems, 86(1), 81–93

Chapter 5
Compensation of Structural Dynamics of Lower Motion Simulator

5.1 Introduction

To ensure experimental accuracy, the design of the HIL simulator is very important. A highly accurate HIL simulation requires that the motion simulator should have high positioning accuracy and high bandwidth frequency response. Therefore, from a mechanical point of view, the motion simulator is expected to have a high structural stiffness and high structural natural frequency. Because parallel-mechanism robots have advantages in structural stiffness and natural frequency, parallel-mechanism robots are often used to design motion simulators (Ananthakrishnan et al.1996; Motaghedi and Stamm, 2005; Osaki et al. 2010; Abiko et al. 2014; Golubev and Yaskevich, 2020; Insam et al. 2021). One problem is that a high structural stiffness leads to high inertia, while a high structural natural frequency requires low inertia. Therefore, the structural stiffness and structural natural frequency are limited and balanced. How the structural dynamics of the motion simulator affect the HIL simulation accuracy should be studied. If the effect is significant, then an error compensation method should be used to improve the experimental accuracy.

In fact, the structural dynamics of the lower motion simulator have non-negligible effects on the accuracy of the HIL simulation of the space docking. Studies (Qi et al. 2018, 2021) showed that the structural dynamics of the upper platform lead to a convergence error in the HIL simulation. An error compensation method based on the upper structural dynamics is studied. For the structural dynamics of the lower motion simulator (i.e., the lower platform), a study (Gao et al. 2016) showed that it leads to a convergence error of the HIL simulation, and an error compensation method based on the lower structural dynamics is used. The study (Qi et al. 2022) found that the structural dynamics of the lower motion simulator can lead to a divergence error of the HIL simulation, and a distributed compensation method is proposed to modulate the motion command of the lower motion simulator. However, how the structural dynamics of the lower motion simulator affect the HIL simulation accuracy is not

C. Qi et al., *Control of Hardware-In-The-Loop Space Contact Simulation Robot*,
https://doi.org/10.1007/978-981-95-1172-3_5

well studied. The conditions of divergence and convergence for the HIL simulation due to the structural dynamics of the lower motion simulator have not been studied.

In this chapter, we consider the effect of structural dynamics of the lower motion simulator on the HIL simulation accuracy and propose a distortion compensation method (Qi et al. 2023). The conditions of the divergence and convergence of the HIL simulation due to the lower motion simulator structural dynamics are given. An error analysis and stability analysis after the distortion compensation are given. Software emulations and experiments are used to verify the analysis and the distortion compensation method. Experiments on real docking mechanisms are conducted to show the applications.

The remaining sections of the chapter are arranged as follows. The modeling of the HIL simulator is given in Sect. 5.2. The effects of system parameters are investigated in Sect. 5.3. Section 5.4 analyses the stability of the HIL simulator. Section 5.5 proposes distortion compensation for the structural dynamics of the lower motion simulator. Some verifications and applications are presented in Sect. 5.6. A summary is given in the last section.

5.2 Modeling of HIL Simulator of Space Docking

Figure 5.1 shows the HIL simulator of the space docking, including a physical simulation subsystem and numerical simulation subsystem.

This chapter mainly focus on the structural dynamics of lower motion simulator, therefore the force measurement delay, dynamic response delay of the motion simulator, the structural dynamics of the upper platform are assumed to be compensated very well.

5.2.1 Structural Dynamics of Lower Motion Simulator

The purpose of the parallel mechanism of lower motion simulator is to realize the relative motion of two spacecraft. In the applications, the motion of the parallel mechanism of lower motion simulator is 6-DOF. The structural dynamics of the lower motion simulator mean the dynamical equation of the lower motion simulator under the desired displacement and the docking force considering the lower motion simulator structural mass, stiffness and damping. The structural dynamics model of the lower motion simulator can be derived from structural mechanics. The equation of a six-dimensional and nonlinear model of structural dynamics is too complicated for a theoretical analysis. At an operating point (posture) in the workspace of the lower motion simulator, a linear 6-DOF structural dynamics model can be obtained using the linearization method. Then a state space MIMO (multi-input multi-output) model can be derived. This MIMO model can be utilized to study the HIL simulation system. However, the analysis based on the MIMO model is also complex.

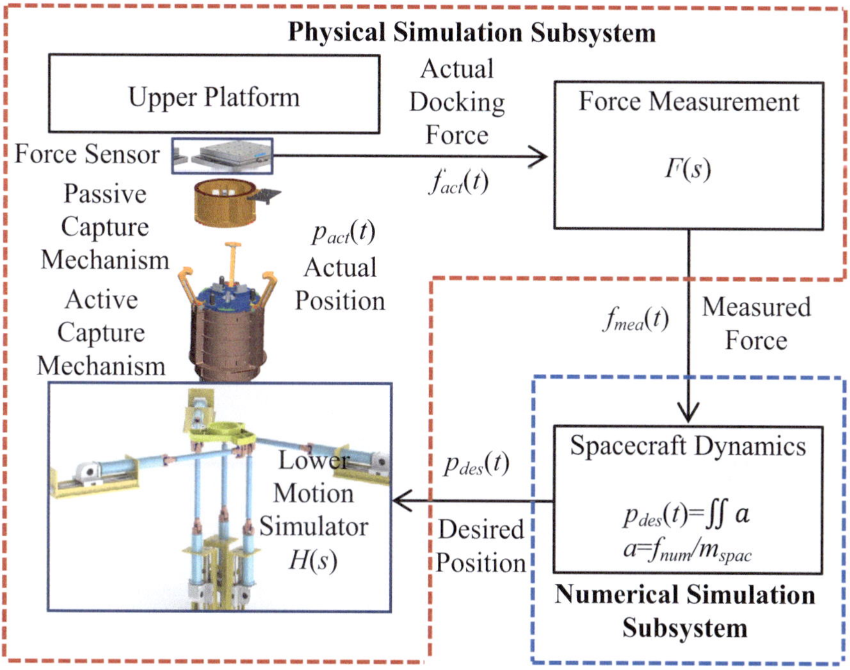

Fig. 5.1 System of HIL simulator of space docking

Therefore, most theoretical studies on HIL simulation systems are based on a one-dimensional model of the HIL simulation system (Osaki et al. 2010; Abiko et al. 2014). In applications, the methods are extended to six-dimensional cases. In the extension to 6-dimensional cases, we assume that there is no/negligible coupling between axes, i.e. their dynamics and control are independent of each other.

The 3–3 orthogonal parallel mechanism has a highly decoupled feature near the orthogonal operating point (i.e., the 3 horizontal links are orthogonal to the 3 vertical links) (Hu et al. 2017). The lower motion simulator in the HIL simulation of the space docking is working near the orthogonal operating point. In the applications, six one-dimensional models with different parameters are reasonable to approximate the structural dynamics of the lower motion simulator. Therefore, to simplify the theoretical analysis, a one-dimensional model along the vertical direction is established as in Fig. 5.2.

Let $f_{act}(t)$ denote the actual contact force on the active docking mechanism, $p_{des}(t)$ denote the desired motion of the lower motion simulator, and $p_{act}(t)$ denote the actual motion of the lower motion simulator. To simplify the modeling, all inertia, stiffness, and damping characteristics of the motion mechanisms are transformed to the moving platform of the lower motion simulator. Then, we can obtain the equivalent mass m_{low}, equivalent structural damping b_{low}, and equivalent structural stiffness k_{low} of the lower motion simulator. Through the force analysis shown in Fig. 5.2, the structural dynamics model of the lower motion simulator can be expressed as

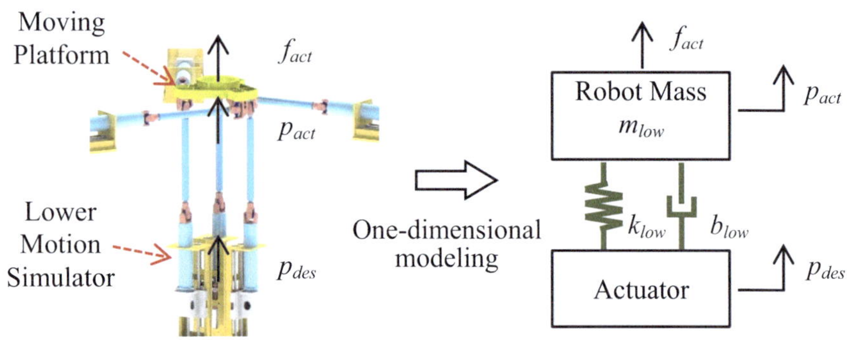

Fig. 5.2 Structural dynamics of lower motion simulator

$$
\begin{aligned}
&m_{low}\frac{d^2(p_{act}(t))}{dt^2} = k_{low}[p_{des}(t) - p_{act}(t)] + \\
&b_{low}\left[\frac{d(p_{des}(t))}{dt} - \frac{d(p_{act}(t))}{dt}\right] + f_{act}(t)
\end{aligned}
\tag{5.1}
$$

It can be rewritten as follows

$$
\begin{aligned}
&m_{low}\frac{d^2(p_{act}(t))}{dt^2} + b_{low}\frac{d(p_{act}(t))}{dt} + k_{low}p_{act}(t) \\
&= b_{low}\frac{d(p_{des}(t))}{dt} + k_{low}p_{des}(t) + f_{act}(t)
\end{aligned}
\tag{5.2}
$$

The above relationship reveals that the actual displacement $p_{act}(t)$ of the lower motion simulator depends on not only the desired displacement $p_{des}(t)$ but also the actual docking force $f_{act}(t)$. Actually, the model 5.2 can be separated into two sub-models as follows

$$
\begin{aligned}
&m_{low}\frac{d^2(\overline{p}_{act}(t))}{dt^2} + b_{low}\frac{d(\overline{p}_{act}(t))}{dt} + k_{low}\overline{p}_{act}(t) = \\
&b_{low}\frac{d(p_{des}(t))}{dt} + k_{low}p_{des}(t)
\end{aligned}
\tag{5.3}
$$

$$
m_{low}\frac{d^2(\Delta p_{act}(t))}{dt^2} + b_{low}\frac{d(\Delta p_{act}(t))}{dt} + k_{low}\Delta p_{act}(t) = f_{act}(t) \tag{5.4}
$$

$$
p_{act}(t) = \overline{p}_{act}(t) + \Delta p_{act}(t) \tag{5.5}
$$

where $\overline{p}_{act}(t)$ denotes the actual motion part of the lower motion simulator without the docking force, $\Delta p_{act}(t)$ denotes the actual additional motion part of the lower motion simulator due to the docking force.

In the complex domain, model 5.2 becomes

$$P_{act}(s) = \frac{b_{low}s + k_{low}}{m_{low}s^2 + b_{low}s + k_{low}} P_{des}(s) + \frac{1}{m_{low}s^2 + b_{low}s + k_{low}} F_{doc}(s) \tag{5.6}$$

From model 5.6, the structural natural frequency can be defined as $f_{low} = \omega_{low}/2\pi$ (Hz), where $\omega_{low} = \sqrt{k_{low}/m_{low}}$ (rad/s), and the structural damping ratio can be defined as $\xi_{low} = b_{low}/(2\sqrt{m_{low}k_{low}})$. The parameters m_{low}, b_{low}, and k_{low} of the above model 5.1 or 5.6 can be initialized from the design software and then calibrated from the calibration experiment.

The dominant structural dynamics of the lower motion simulator should be captured by the simplified model. The model order can be adjusted according to the modeling accuracy. In our applications, a second-order model can capture the dominant structural dynamics of the lower motion simulator. Experimental results show that this model used for the compensator design can ensure the simulation accuracy. More complicated model can also be used to describe the dynamic behavior of the lower motion simulator. It depends on the complexity of the structural dynamics of the lower motion simulator.

5.2.2 *Contact Dynamics*

The real docking mechanisms are six-dimensional contacts, but the contact results are not known, therefore the six-dimensional contact experiments on real docking mechanisms cannot be used to evaluate the accuracy of the HIL simulation. In space, an undamped elastic contact of two objects behaves as a constant amplitude oscillation system. Considering the feasibility of the experiment and the simplicity of the analysis, a one-dimensional undamped elastic contact model between two docking mechanisms is considered:

$$f_{act}(t) = -k_{con}p_{act}(t) \tag{5.7}$$

where k_{con} is the contact stiffness between the active and passive docking mechanisms. The ideal result of a one-dimensional undamped elastic contact in space is known, which can be used to evaluate the accuracy of the HIL simulation. Of course, the model 5.7 can also be changed to include a damping term for energy dissipation, and then the following equations will need some revisions. In this study, the contact stiffness is selected within a range of the real contact stiffness. At the same time, the contact stiffness can be identified in real time from the contact experiment data. Here, the contact stiffness can be time-varying; therefore, model 5.7 can be considered a linearized time-varying model for nonlinear contact dynamics.

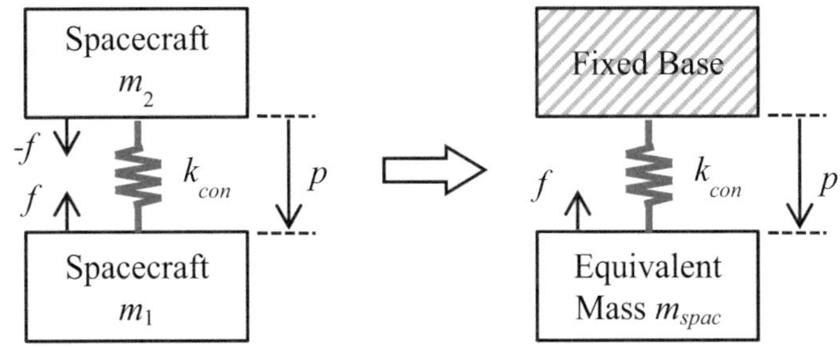

Fig. 5.3 Spacecraft motion model

5.2.3 Spacecraft Motion Model Under Contact Force

The spacecraft motion model receives the docking force and then calculates the relative motion of two spacecraft in zero-gravity space for the next simulation step. Using a high-performance real-time computing module, the model computation time is negligible. It can be proven that the relative motion of two spacecraft with masses m_1 and m_2 equals the motion of an equivalent spacecraft with a mass $m_{spac} = m_1 m_2/(m_1 + m_2)$, as shown in Fig. 5.3. According to Newton's second law,

$$m_{spac} \frac{d^2 p_{des}(t)}{dt^2} = f_{act}(t) \tag{5.8}$$

where m_{spac} is the equivalent mass of two spacecraft. The contact process modeled by Eqs. 5.7 and 5.8 can be considered a mass spring system. Therefore, there is a parameter called the natural frequency $\omega_{con} = \sqrt{k_{con}/m_{spac}}$ (rad/s) or $f_{con} = \omega_{con}/2\pi$ (Hz) to describe the contact characteristic mode.

5.2.4 Model of HIL Simulator of Space Docking

The model of the HIL simulator of space docking consists of the spacecraft motion model, lower motion simulator structural dynamics, and docking mechanism contact dynamics. The model of the HIL simulator of the space docking is constructed in Fig. 5.4.

The contact process begins when the active docking mechanism on the active spacecraft impacts the passive docking mechanism on the passive spacecraft at an initial relative velocity v_{init}. The initial relative velocity v_{init} of two spacecraft is one parameter of the docking initial conditions. The initial relative position of the two spacecraft is set to zero without loss of generality. The desired velocity of the lower motion simulator is denoted by $v_{des}(t)$. For the whole model of the HIL simulator, the input is the initial relative velocity v_{init}, and the output is the desired velocity of the lower motion simulator. Of course, the actual docking force $f_{act}(t)$ on the active

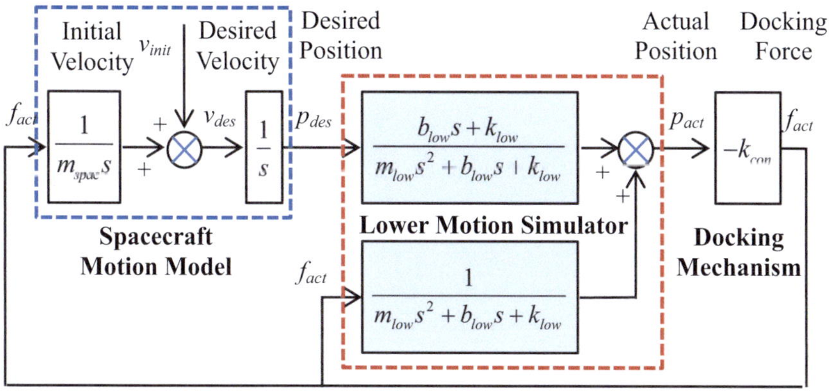

Fig. 5.4 Block diagram of HIL simulator of space docking

docking mechanism can also be considered as the output. The whole closed-loop model of the HIL simulator can be obtained as

$$G_{\text{simulator}}^{\text{close}}(s) = \frac{V_{des}(s)}{V_{init}(s)}$$
$$= \frac{m_{spac}s^2(m_{low}s^2 + b_{low}s + k_{low} + k_{con})}{m_{spac}s^2(m_{low}s^2 + b_{low}s + k_{low} + k_{con}) + k_{con}(b_{low}s + k_{low})} \quad (5.9)$$

using closed-loop control theory (Ogata 2009). The model 5.9 will be used for the following distortion effect analysis and stability analysis.

5.3 Effect of System Parameters on Simulation Distortion

In space, the motion of two spacecraft for an ideal elastic contact is a simple harmonic motion (SHM) with a natural frequency f_{con}, which is considered as an ideal case. For the HIL simulation, two spacecraft may not move like an SHM due to the structural dynamics of the lower motion simulator. To investigate the effect of the structural dynamics of the lower motion simulator on the accuracy of the HIL simulation, some typical parameters are selected as an example. These parameters are within a reasonable range for a real HIL simulator and a real space docking. Of course, other reasonable parameter values can also be used. However, they do not change the trends and conclusions. For the lower motion simulator, the equivalent mass m_{low} is 100 kg, the structural natural frequency f_{low} is 30 Hz, and the structural damping ratio ξ_{low} is 0.8. For the spacecraft and the docking mechanism, the equivalent mass m_{spac} of two spacecraft is 50 kg, and the natural frequency f_{con} of the contact dynamics is 8 Hz. Note that the equivalent mass m_{spac} and contact stiffness k_{con} between two docking mechanisms determine the natural frequency f_{con} of the contact dynamics. The initial

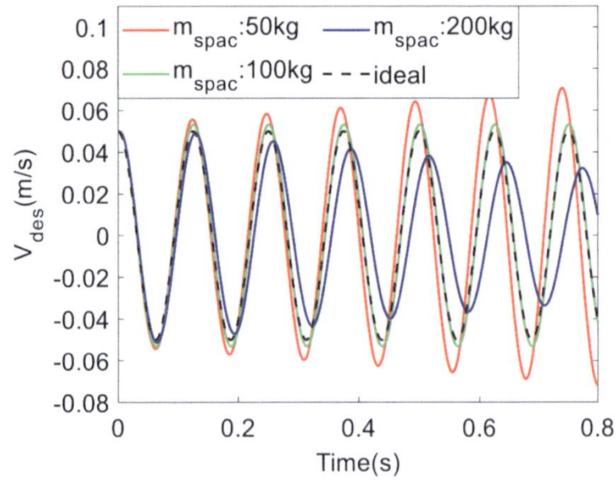

Fig. 5.5 Desired relative velocity $v_{des}(t)$ of two spacecraft simulated under different equivalent masses m_{spac} of two spacecraft (m_{low}: 100 kg, f_{low}: 30 Hz, ξ_{low}: 0.8, m_{spac}: 50 kg, 100 kg, 200 kg, f_{con}: 8 Hz, v_{init}: 0.05 m/s)

docking velocity v_{init} (i.e., the docking initial condition) of the two spacecraft is 0.05 m/s. By running the HIL simulator model as shown in Fig. 5.4, we can obtain the simulation results. In order to save the space, only a few sets of simulation results under some selected parameters are given. One or two parameters will change in order to show the effects of parameters on the simulation distortion.

5.3.1 Effect of Mass of Spacecraft

Figure 5.5 exhibits the desired relative velocity $v_{des}(t)$ of two spacecraft simulated under different equivalent masses m_{spac} of the two spacecraft. When the equivalent mass m_{spac} of two spacecraft is 50 kg, 100 kg, and 200 kg, compared with the ideal case, the spacecraft move in divergent motion, SHM, and convergent motion, respectively. It shows that whether the system is convergent or divergent depends on the relative value of m_{spac} and m_{low}. This phenomenon is undesirable because the accuracy of the HIL simulation should not be related to the equivalent mass of two spacecraft. In the following, a divergent case is selected as the baseline of the comparison, where m_{spac} is 50 kg, m_{low} is 100 kg, f_{low} is 30 Hz, ξ_{low} is 0.8, f_{con} is 8 Hz, and v_{init} is 0.05 m/s.

5.3.2 Effect of Natural Frequency of Contact Dynamics

Figure 5.6 exhibits the desired relative velocity $v_{des}(t)$ of two spacecraft simulated under different natural frequencies f_{con} of the contact dynamics. The natural frequency of contact dynamics is changed with different contact stiffness since the equivalent mass of two spacecraft is fixed. It shows that for a higher natural frequency

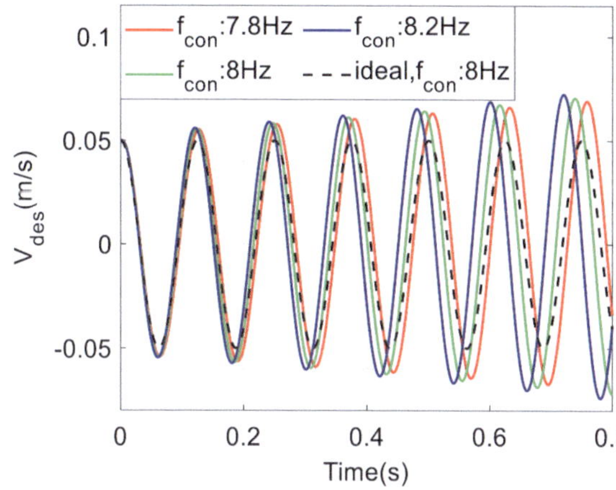

Fig. 5.6 Desired relative velocity $v_{des}(t)$ of two spacecraft simulated under different natural frequencies f_{con} of contact dynamics (m_{low}: 100 kg, f_{low}: 30 Hz, ξ_{low}: 0.8, m_{spac}: 50 kg, f_{con}: 7.8 Hz, 8.0 Hz, 8.2 Hz, v_{init}: 0.05 m/s)

f_{con} of the contact dynamics, the divergence of the HIL simulation is more significant for divergent cases (CoR > 1).

The distortion level (i.e., the convergence or divergence level) of the HIL simulation system can be measured by the coefficient of restitution (CoR). The CoR is calculated as the average ratio between two adjacent peaks of the desired relative velocity curve of two spacecraft in a certain simulation time. The CoR of a convergent system is lower than 1, while the CoR of a divergent system is higher than 1. The farther the CoR is from 1, the more serious the distortion of the HIL simulation system.

Figure 5.7 shows the CoR under different natural frequencies of contact dynamics with three cases of the equivalent mass of two spacecraft. The natural frequency of contact dynamics is changed with different contact stiffness when the equivalent mass of two spacecraft is fixed. Whether the system is convergent or divergent depends on the relative values of m_{spac} and m_{low}. When the natural frequency of contact dynamics grows, the CoR (red line and green line) deviates more from the ideal value (blue line, CoR: 1).

5.3.3 *Effect of Structural Natural Frequency and Damping Ratio of Lower Motion Simulator*

Figure 5.8 exhibits the desired relative velocity $v_{des}(t)$ of two spacecraft simulated under different structural natural frequencies f_{low} of the lower motion simulator. It shows that as the structural natural frequency f_{low} decreases, the divergence of the HIL simulation becomes more significant.

Figure 5.9 shows the CoR under different structural natural frequencies of the lower motion simulator with three cases of the equivalent mass of two spacecraft. When the structural natural frequency of the lower motion simulator declines, the

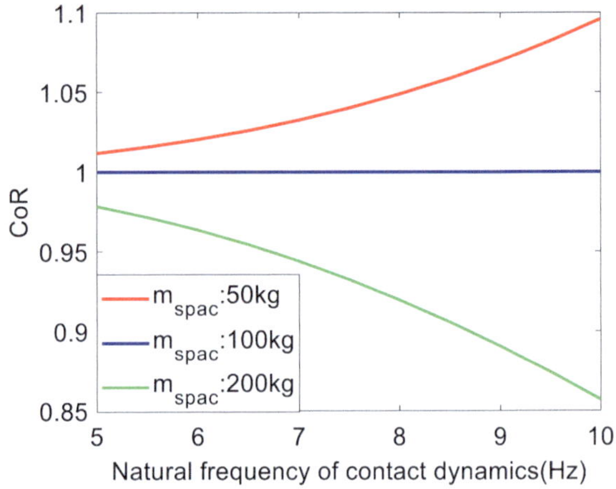

Fig. 5.7 CoR at different natural frequencies f_{con} of contact dynamics (m_{low}: 100 kg, f_{low}: 30 Hz, ξ_{low}: 0.8, m_{spac}: 50 kg, 100 kg, 200 kg, f_{con}: 5 Hz ~ 10 Hz, v_{init}: 0.05 m/s)

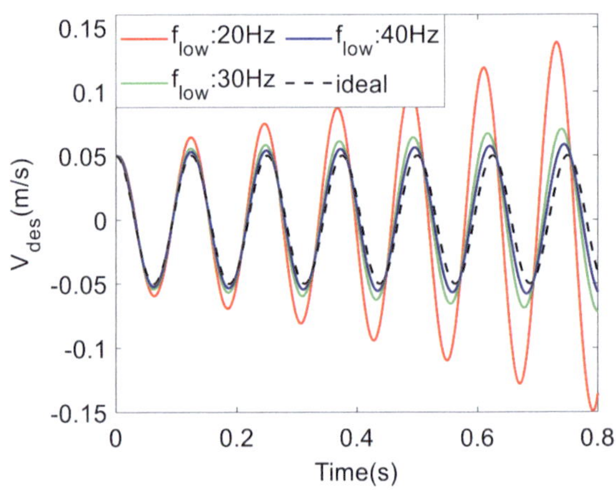

Fig. 5.8 Desired relative velocity $v_{des}(t)$ of two spacecraft simulated under different structural natural frequencies f_{low} of lower motion simulator (m_{low}: 100 kg, f_{low}: 20 Hz, 30 Hz, 40 Hz, ξ_{low}: 0.8, m_{spac}: 50 kg, f_{con}: 8 Hz, v_{init}: 0.05 m/s)

CoR (red line and green line) deviates more from the ideal value (blue line, CoR: 1). Although a higher structural natural frequency of the lower motion simulator reduces the simulation distortion, a high structural natural frequency of the lower motion simulator is very difficult to achieve in practice.

Figure 5.10 exhibits the desired relative velocity $v_{des}(t)$ of two spacecraft simulated under different structural damping ratios ξ_{low} of the lower motion simulator. As the structural damping ratio ξ_{low} of the lower motion simulator increases, the divergence of the HIL simulation becomes more significant.

Figure 5.11 shows the CoR under different structural damping ratios of the lower motion simulator with three cases of the equivalent mass of two spacecraft. When the structural damping ratio grows, the CoR (red line and green line) deviates more from the ideal value (blue line, CoR: 1). This means that the system suffers from more serious distortion at a higher structural damping ratio of the lower motion simulator,

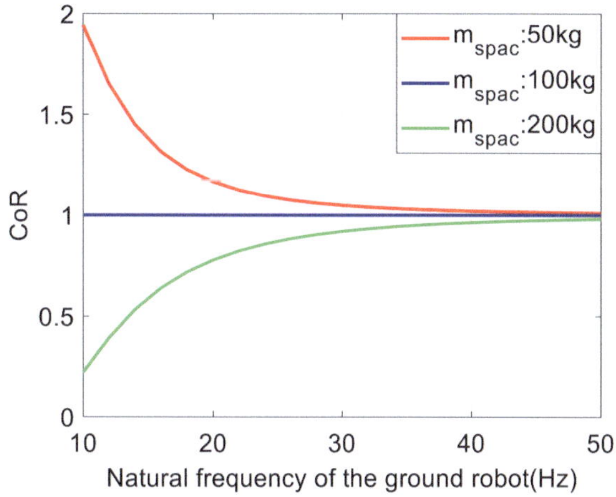

Fig. 5.9 CoR under different structural natural frequencies f_{low} of lower motion simulator (m_{low}: 100 kg, f_{low}: 10 Hz ~ 50 Hz, ξ_{low}: 0.8, m_{spac}: 50 kg, 100 kg, 200 kg, f_{con}: 8 Hz, v_{init}: 0.05 m/s)

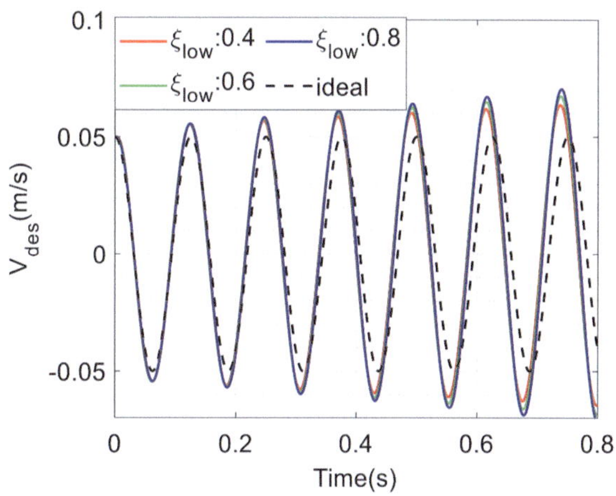

Fig. 5.10 Desired relative velocity $v_{des}(t)$ of two spacecraft simulated under different structural damping ratios ξ_{low} of lower motion simulator (m_{low}: 100 kg, f_{low}: 30 Hz, ξ_{low}: 0.4, 0.6, 0.8, m_{spac}: 50 kg, f_{con}: 8 Hz, v_{init}: 0.05 m/s)

but the effect is relatively slight compared with that of the structural natural frequency of the lower motion simulator.

In summary, whether the system is convergent or divergent depends on the relative value of m_{spac} and m_{low}. For a higher natural frequency f_{con} of the contact dynamics, a lower structural natural frequency f_{low} of the lower motion simulator, and a higher structural damping ratio ξ_{low} of the lower motion simulator, the divergence of the HIL simulation is more significant for divergent cases (CoR > 1). Similarly, the convergence of the HIL simulation is more significant for convergent cases (CoR < 1).

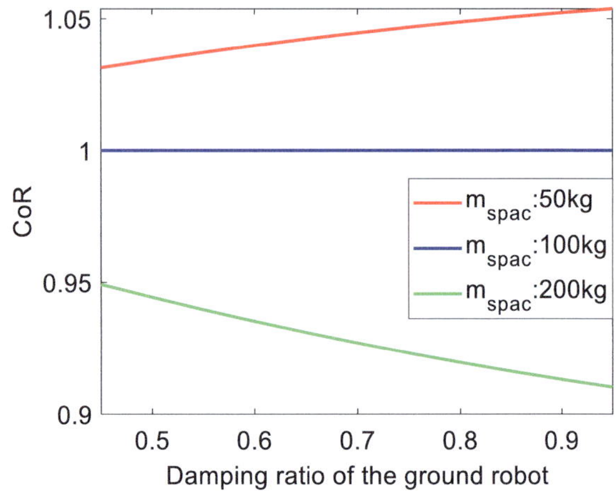

Fig. 5.11 CoR under different structural damping ratios ξ_{low} of lower motion simulator (m_{low}: 100 kg, f_{low}: 30 Hz, ξ_{low}: 0.45 ~ 0.95, m_{spac}: 50 kg, 100 kg, 200 kg, f_{con}: 8 Hz, v_{init}: 0.05 m/s)

5.4 Stability Analysis of HIL Simulator

5.4.1 Stability Conditions

A stability analysis of the HIL simulator of the space docking can explain why the structural dynamics of the lower motion simulator leads to HIL simulation divergence or convergence and provides relevant critical conditions.

To obtain the stability conditions of the closed-loop system 5.9 of the HIL simulator, the relative stability is analyzed from the perspective of the gain margin and phase margin. Note that the feedback loop in Fig. 5.4 is positive; then, the transfer function of the open-loop system is

$$G_{without\_comp}^{open}(s) = \frac{k_{con}(b_{low}s + k_{low})}{m_{spac}s^2(m_{low}s^2 + b_{low}s + k_{low} + k_{con})} \tag{5.10}$$

Correspondingly, the frequency–response characteristics of the open-loop system can be expressed as

$$A(\omega) = \frac{k_{con}}{m_{spac}\omega^2}\sqrt{\frac{k_{low}^2 + b_{low}^2\omega^2}{(k_{low} + k_{con} - m_{low}\omega^2)^2 + b_{low}^2\omega^2}} \tag{5.11}$$

$$\Phi(\omega) = \arctan\frac{b_{low}\omega}{k_{low}} - \arctan\frac{b_{low}\omega}{k_{low} + k_{con} - m_{low}\omega^2} - \pi \tag{5.12}$$

where $A(\omega)$ and $\Phi(\omega)$ are the magnitude ratio and phase angle, respectively.

Let ω_c denote the gain-crossover frequency and ω_g denote the phase-crossover frequency (Ogata 2009). From

$$A(\omega_c) = 1 \tag{5.13}$$

a high-order equation for the gain-crossover frequency ω_c can be obtained. This shows that the gain-crossover frequency is related to the characteristics of the lower motion simulator and the spacecraft, and the equation is difficult to solve analytically. Therefore, the gain-crossover frequency ω_c is calculated numerically from Eq. 5.13. Then, the phase margin $\gamma(\omega_c)$ is obtained as follows (Ogata 2009):

$$\gamma(\omega_c) = \pi + \Phi(\omega_c) \tag{5.14}$$

From

$$\pi + \Phi(\omega_g) = 0 \tag{5.15}$$

the phase-crossover frequency can be obtained as

$$\omega_g = \sqrt{k_{con}/m_{low}} \tag{5.16}$$

Then, the gain margin can be calculated by (Ogata, 2009)

$$K_g = \frac{1}{A(\omega_g)} = \frac{m_{spac}}{m_{low}} \tag{5.17}$$

From the gain margin 5.17, the stability conditions are presented below.

- When the equivalent mass m_{spac} of two spacecraft equals the equivalent mass m_{low} of the lower motion simulator (i.e., $m_{spac} = m_{low}$), the gain margin equals 1 (i.e., $K_g = 1$). In this case, the HIL simulation system is marginally stable, and the simulated relative motion of two spacecraft is an SHM.
- When $m_{spac} > m_{low}$, the gain margin is higher than 1 (i.e., $K_g > 1$). In this case, the HIL simulation system is asymptotically stable, and the simulated relative motion of the two spacecraft is convergent motion.
- When $m_{spac} < m_{low}$, the gain margin is lower than 1 (i.e., $K_g < 1$). In this case, the HIL simulation system is unstable, and the simulated relative motion of the two spacecraft is divergent motion.

This means that, considering the structural dynamics of the lower motion simulator, the stability of the HIL simulator is determined by the ratio of the equivalent mass of two spacecraft to the mass of the lower motion simulator. The divergent, SHM, and convergent results in Fig. 5.5 under different equivalent masses of the two spacecraft coincide with the derived stability conditions.

5.4.2 Phase Margin Analysis with Mass of Spacecraft

To further study how the distortion of the HIL simulation system varies with the equivalent mass of two spacecraft, the phase margin under different equivalent masses of two spacecraft is calculated from 5.14. The equivalent mass m_{low} of the lower motion simulator is 100 kg, the structural natural frequency f_{low} of the lower motion simulator is 30 Hz, the structural damping ratio ξ_{low} of the lower motion simulator is 0.8, and the natural frequency f_{con} of the contact dynamics is 8 Hz.

As shown in Fig. 5.12, the horizontal black dashed line represents a marginally stable line with a zero phase margin. This line intersects with the phase margin line when $m_{spac} = m_{low}$. Above the marginally stable line, the HIL simulation system is stable. Below the marginally stable line, the HIL simulation system is unstable. The convergence or divergence distortion arises when the state of the HIL simulation system deviates from the marginally stable line. In the stable area, as m_{spac} increases, the phase margin also increases, and the rate of the phase margin decreases. Therefore, with increasing m_{spac} in the stable area, the convergence distortion of the HIL simulation becomes more serious, and the rate of convergence distortion decreases. In the unstable area, as m_{spac} decreases, the phase margin also decreases, and the rate of the phase margin increases. Therefore, with a decrease in m_{spac} in the unstable area, the divergence distortion of the HIL simulation becomes more serious, and the rate of divergence distortion increases.

The relation curve between the distortion level of the HIL simulation and the equivalent mass of two spacecraft is demonstrated by Fig. 5.13. This illustrates that with increasing m_{spac} in the stable area, the CoR decreases and the decreasing rate of the CoR decreases. With a decrease in m_{spac} in the unstable area, the CoR increases, and the increasing rate of the CoR increases. This result is in good agreement with the distortion analysis from the phase margin in Fig. 5.12.

Fig. 5.12 Phase margin under different equivalent masses of two spacecraft

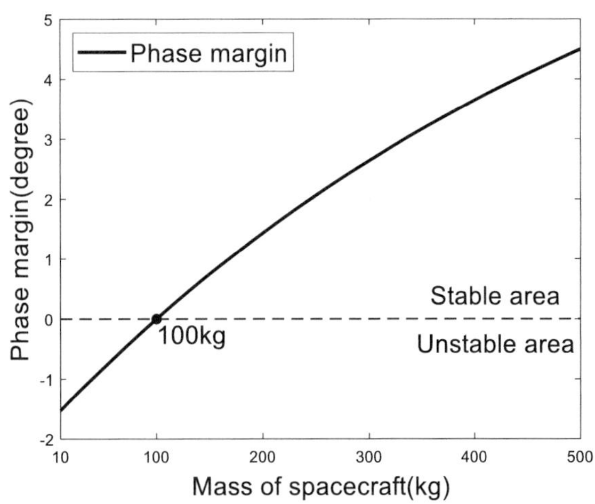

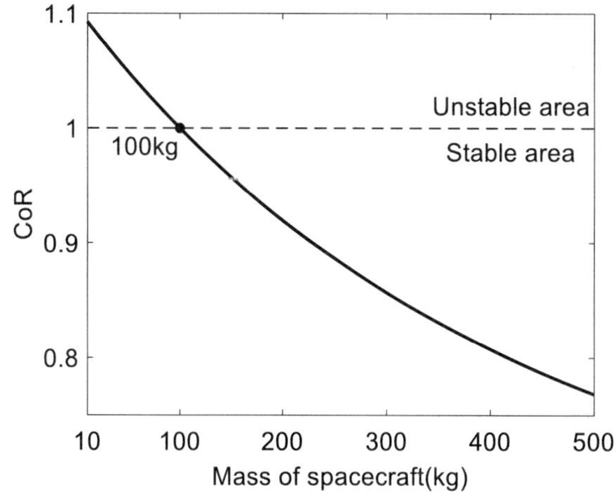

Fig. 5.13 CoR under different equivalent masses of the two spacecraft

5.4.3 Phase Margin Analysis with Natural Frequency of Contact Dynamics

Given the mass of the lower motion simulator and the equivalent mass of the two spacecraft, other parameters have additional effects on the distortion level of the HIL simulation. Figure 5.14 shows the phase margin under different natural frequencies of contact dynamics with the same parameters as Fig. 5.7. The natural frequency of contact dynamics is changed with different contact stiffness when the equivalent mass of two spacecraft is fixed. Whether the phase margin is positive or negative depends on the relative value of m_{spac} and m_{low}. When the natural frequency of contact dynamics grows, the phase margin deviates more from the marginally stable situation (blue line, CoR: 1), and the system distortion is aggravated. This is the reason for the simulation result in Fig. 5.7.

5.4.4 Phase Margin Analysis with Structural Natural Frequency and Damping Ratio of Lower Motion Simulator

Figure 5.15 shows the phase margin under different structural natural frequencies of the lower motion simulator with the same parameters as Fig. 5.9. When the structural natural frequency of the lower motion simulator declines, the phase margin deviates more from the marginally stable situation (blue line, CoR: 1). This is the reason for the simulation result in Fig. 5.9.

Figure 5.16 shows the phase margin under different structural damping ratios of the lower motion simulator with the same parameters as Fig. 5.11. When the structural damping ratio grows, the CoR and the phase margin deviate more from the

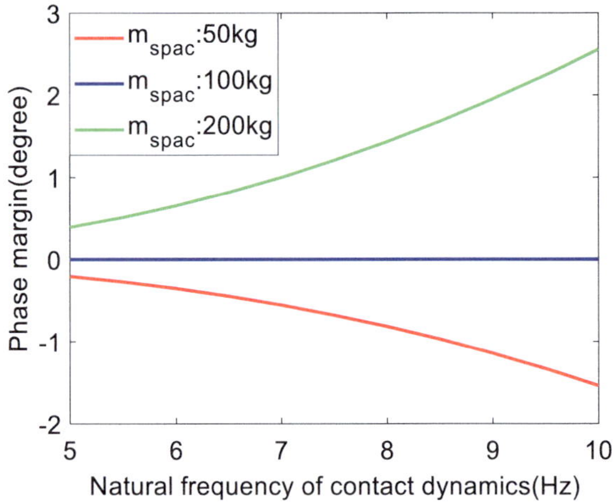

Fig. 5.14 Phase margin at different natural frequencies f_{con} of contact dynamics (m_{low}: 100 kg, f_{low}: 30 Hz, ξ_{low}: 0.8, m_{spac}: 50 kg, 100 kg, 200 kg, f_{con}: 5 Hz ~ 10 Hz, v_{init}: 0.05 m/s)

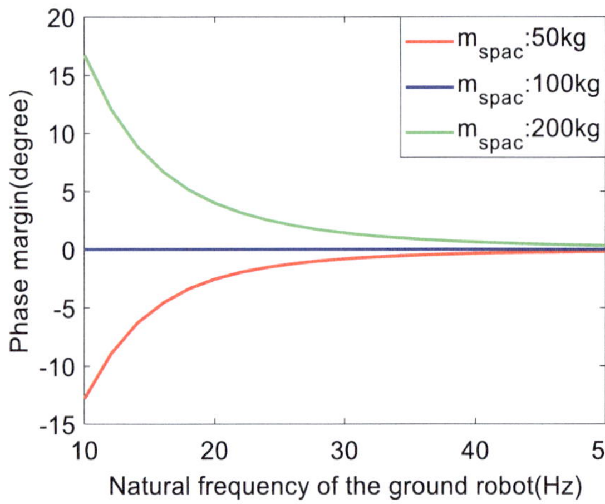

Fig. 5.15 Phase margin at different structural natural frequencies f_{low} of lower motion simulator (m_{low}: 100 kg, f_{low}: 10 Hz ~ 50 Hz, ξ_{low}: 0.8, m_{spac}: 50 kg, 100 kg, 200 kg, f_{con}: 8 Hz, v_{init}: 0.05 m/s)

marginally stable situation (blue line, CoR: 1). This is the reason for the simulation result in Fig. 5.11.

To ensure the stability and accuracy of the HIL simulation, a reasonable CoR should be slightly lower than 1. This requires the equivalent mass of two spacecraft to be slightly higher than the mass of the lower motion simulator. However, the HIL simulator of space docking should be able to simulate various spacecraft; thus, the method based on the design of the mass ratio is not a good method. Therefore, some new effective compensation methods for the structural dynamics of the lower motion simulator are needed.

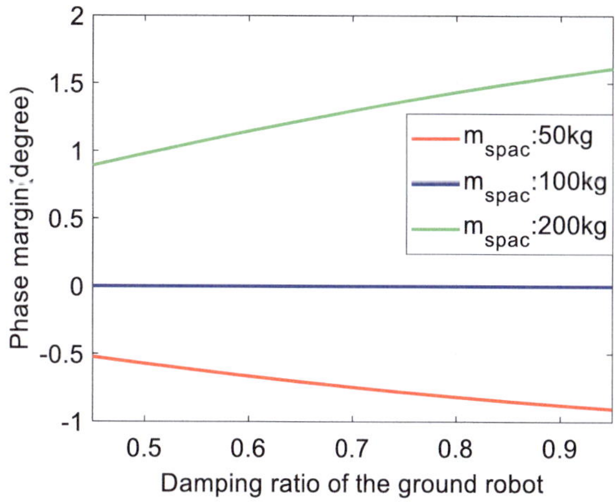

Fig. 5.16 Phase margin under different structural damping ratios ξ_{low} of lower motion simulator (m_{low}: 100 kg, f_{low}: 30 Hz, ξ_{low}: 0.45 ~ 0.95, m_{spac}: 50 kg, 100 kg, 200 kg, f_{con}: 8 Hz, v_{init}: 0.05 m/s)

5.5 Distortion Compensation for Structural Dynamics of Lower Motion Simulator

5.5.1 *The Proposed Distortion Compensation Method*

The structural dynamics of the lower motion simulator bring two distortion factors into the HIL simulator. One is a delay block, and the other is a lead block, as shown in Fig. 5.17. Let $\overline{p}_{act}$ denote the output of the delay block, and then the transfer functions of two blocks can be expressed as follows:

$$D(s) = \frac{\overline{P}_{act}(s)}{P_{des}(s)} = \frac{b_{low}s + k_{low}}{m_{low}s^2 + b_{low}s + k_{low}} \tag{5.18}$$

$$L(s) = \frac{F_{act}(s)}{\overline{P}_{act}(s)} = -k_{con}\frac{m_{low}s^2 + b_{low}s + k_{low}}{m_{low}s^2 + b_{low}s + k_{low} + k_{con}} \tag{5.19}$$

The delay block 5.18 represents the delay from the desired motion of the lower motion simulator to the actual motion of the lower motion simulator without considering the docking force, whose time domain model is Eq. 5.3. The phase angle is dependent on the lower motion simulator structural parameters m_{low}, k_{low}, b_{low} and the system frequency ω as follows

$$\Phi_D(\omega) = \arctan\frac{b_{low}\omega}{k_{low}} - \arctan\frac{b_{low}\omega}{k_{low} - m_{low}\omega^2} \tag{5.20}$$

This phase angle is always negative which means that this block is a delay block. The delay time in the time domain can be derived from $\Phi_D(\omega)/\omega$. Note that this delay is dynamic which is different from the pure delay time $e^{-t_d s}$.

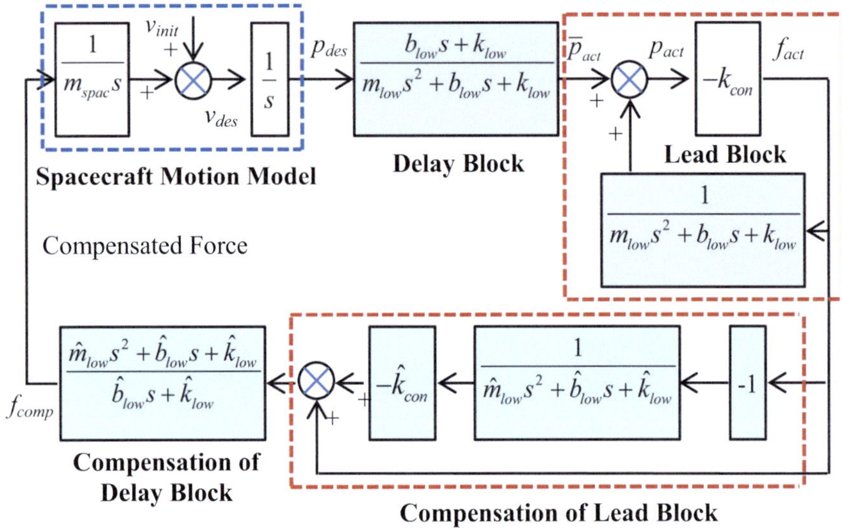

Fig. 5.17 The proposed distortion compensation method

The lead block 5.19 represents the lead from the actual motion of the lower motion simulator without considering the docking force to the actual docking force, whose complex domain model can be derived from Fig. 5.17 and whose time domain model can be derived from Eqs. 5.4 and 5.7. Its phase angle is derived as follows

$$\Phi_L(\omega) = \arctan\frac{b_{low}\omega}{k_{low} - m_{low}\omega^2} - \arctan\frac{b_{low}\omega}{k_{low} + k_{con} - m_{low}\omega^2} \tag{5.21}$$

This phase angle is always positive which means that this block is a lead block. The lead time in the time domain can be derived from $\Phi_L(\omega)/\omega$.

Due to the above two blocks, the contact force simulated by the HIL simulator is different from that of the docking system in space. The principle of distortion compensation is to make the contact force from the ground simulation the same as the real force in space. The proposed compensation model is shown in Fig. 5.17, where two distortion factors are compensated individually by two phase shift models.

The lead block $L(s)$ is compensated by a delay compensator

$$L_{comp}(s) = 1 + \frac{\hat{k}_{con}}{\hat{m}_{low}s^2 + \hat{b}_{low}s + \hat{k}_{low}} \tag{5.22}$$

so that $L(s)L_{comp}(s)$ approaches $-k_{con}$, where $\hat{k}_{con}$ is the identified contact stiffness and $\hat{m}_{low}$, $\hat{b}_{low}$, and $\hat{k}_{low}$ are the parameters of the compensator for the structural dynamics. The compensation parameters of the structural dynamics can be identified by offline experiments.

The delay block $D(s)$ is compensated by a lead compensator

$$D_{comp}(s) = \frac{\hat{m}_{low}s^2 + \hat{b}_{low}s + \hat{k}_{low}}{\hat{b}_{low}s + \hat{k}_{low}} \tag{5.23}$$

so that $D(s)D_{comp}(s)$ approximates 1.

It can be assumed that the contact stiffness remains constant in several milliseconds during the space docking process. Therefore, the contact stiffness can be identified from the contact force and motion based on the following equation:

$$\Delta f_{act}(t) = \hat{k}_{con}(t)\Delta p_{act}(t) \tag{5.24}$$

where $\Delta f_{act}(t) = f_{act}(t) - f_{act}(t-1)$ and $\Delta p_{act}(t) = p_{act}(t) - p_{act}(t-1)$. To resist noise jamming, in this study, $\hat{k}_{con}$ is identified by the least square method based on the model 5.24 from the data $\{\Delta f_{act}(t), ..., \Delta f_{act}(t-n), \Delta p_{act}(t), ..., \Delta p_{act}(t-n)\}$, where n means the past n moments.

When the real docking mechanisms have energy loss, but the contact dynamics modeling do not consider the energy loss (i.e., without damping term), then the simulation could have a little convergence distortion. In practice, considering the stability and safety, a little convergence error is expected. If the energy loss of the docking mechanisms is very large, the damping term should be added in the contact dynamics modeling and identification.

5.5.2 Stability Analyses of HIL Simulator After Distortion Compensation

The error of the HIL simulation is analyzed by comparing the HIL simulation system with the docking system in space. For the HIL simulation system, the initial relative velocity v_{init} of the two spacecraft is considered as the model input, and the desired relative velocity $v_{des}(t)$ of the two spacecraft is the model output. From Fig. 5.17, the closed-loop model of the HIL simulation system after compensation can be expressed by

$$\begin{aligned} G_{comp}^{close}(s) &= \frac{V_{des}(s)}{V_{init}(s)} \\ &= \frac{m_{spac}s^2}{m_{spac}s^2 + k_{con}\frac{(b_{low}s+k_{low})(\hat{m}_{low}s^2+\hat{b}_{low}s+\hat{k}_{low}+\hat{k}_{con})}{(\hat{b}_{low}s+\hat{k}_{low})(m_{low}s^2+b_{low}s+k_{low}+k_{con})}} \end{aligned} \tag{5.25}$$

The closed-loop transfer function of the docking system in space is

$$G_{space}^{close}(s) = \frac{V_{des}(s)}{V_{init}(s)} = \frac{m_{spac}s^2}{m_{spac}s^2 + k_{con}} \tag{5.26}$$

When the parameters of the compensator are matched with those of the lower motion simulator, i.e., $\hat{m}_{low} = m_{low}$, $\hat{b}_{low} = b_{low}$, $\hat{k}_{low} = k_{low}$, and the identified contact stiffness is matched with the real contact stiffness as well, i.e., $\hat{k}_{con} = k_{con}$, the closed-loop model of the HIL simulation system after the compensation becomes $G^{close}_{comp\_ideal}(s) = G^{close}_{space}(s)$. In this case, the error of the HIL simulation is $E(s) = G^{close}_{comp\_ideal}(s) - G^{close}_{space}(s) = 0$. This indicates that the HIL simulation system can provide results with high accuracy regardless of the equivalent mass of the two spacecraft and the contact dynamics.

In the implementation, the parameters of the compensator can be matched with the lower motion simulator, i.e., $\hat{m}_{low} = m_{low}$, $\hat{b}_{low} = b_{low}$, $\hat{k}_{low} = k_{low}$, through the offline identification of the structural dynamics of the lower motion simulator. However, due to the complicated contact situation and the measurement noise, it is difficult to match the identified contact stiffness with the real contact stiffness online, i.e., $\hat{k}_{con} \neq k_{con}$. Let $d_k = (\hat{k}_{con} - k_{con})/k_{con}$ denote the identification error of the contact stiffness. Then, the closed-loop model of the HIL simulation system after compensation can be expressed by

$$\begin{aligned} &G^{close}_{comp\_with\_stiffness\_error}(s) \\ &= \frac{m_{spac}s^2}{m_{spac}s^2 + k_{con}(1 + d_k \frac{k_{con}}{m_{low}s^2 + b_{low}s + k_{low} + k_{con}})} \end{aligned} \tag{5.27}$$

To analyze the stability of the HIL simulation system with the contact stiffness identification error, the phase margin of the open-loop system

$$\begin{aligned} &G^{open}_{comp\_with\_stiffness\_error}(s) \\ &= \frac{k_{con}(m_{low}s^2 + b_{low}s + k_{low} + k_{con} + d_k k_{con})}{m_{spac}s^2(m_{low}s^2 + b_{low}s + k_{low} + k_{con})} \end{aligned} \tag{5.28}$$

can be calculated as

$$\begin{aligned} &\phi_{comp\_with\_stiffness\_error}(\omega) \\ &= \arg\tan \frac{b_{low}\omega}{k_{low} + (1 + d_k)k_{con} - m_{low}\omega^2} \\ &- \arg\tan \frac{b_{low}\omega}{k_{low} + k_{con} - m_{low}\omega^2} \end{aligned} \tag{5.29}$$

It can be shown that when d_k is a positive constant, the phase margin is lower than zero, and then the HIL simulation system is unstable. Because the HIL simulation system should remain stable, the identified contact stiffness $\hat{k}_{con}$ must be lower than the actual contact stiffness k_{con}, i.e.,

$$\hat{k}_{con} < k_{con} \tag{5.30}$$

In practice, d_k is usually a random disturbance. In this case, the mean value of $\hat{k}_{con}$ is required to be lower than that of k_{con}, i.e.,

$$E[\hat{k}_{con}] < k_{con} \tag{5.31}$$

in order to guarantee the system stability.

5.6 Verifications and Applications

5.6.1 Verifications by Software

The effectiveness of the proposed compensation method is evaluated through a software emulation of the HIL simulation system. In the software emulation, typical parameters are selected. For the lower motion simulator, m_{low} is 100 kg, f_{low} is 30 Hz, and ξ_{low} is 0.8. The initial docking velocity v_{init} is 0.05 m/s. For the spacecraft and the docking mechanism, the equivalent mass m_{spac} has two values, 50 kg or 200 kg, which lead to divergence or convergence distortion, respectively. The natural frequency f_{con} of the contact dynamics has two values: 5 Hz or 8 Hz. Note that the conclusions are the same when adopting different reasonable parameter settings.

Figure 5.18 exhibits the desired relative velocity $v_{des}(t)$ of two spacecraft with $m_{spac} = 50$ kg and $f_{con} = 8$ Hz. The curves with the legend of uncompensated, lead-block-compensated, delay-block-compensated, fully compensated, and ideal results represent the HIL simulation without compensation, only compensating the lead block, only compensating the delay block, with the proposed full compensation (compensating both the delay block and lead block) and under ideal conditions, respectively. If the HIL simulation system is not compensated, then divergence distortion arises (red line, CoR: 1.0487). Only compensating for the lead block makes the divergence distortion more serious (blue line, CoR: 1.1043). Only compensating for the delay block generates serious convergence distortion (magenta line, CoR: 0.9572). After full compensation, the HIL simulation has high accuracy (green line, CoR: 1.0000).

Figure 5.19 display the compensation performance for the convergence distortion case with $m_{spac} = 200$ kg, $f_{con} = 8$ Hz. The HIL simulation system is convergent without compensation (red line, CoR: 0.9195). With the compensation for the lead block, the convergence distortion becomes a divergence distortion (blue line, CoR: 1.1043). With compensation for the delay block, the convergence distortion worsens (magenta line, CoR: 0.8568). Under the proposed full compensation, the simulation results (green line, CoR: 1.0000) are very close to the ideal cases.

Table 5.1 summarizes the CoR of the results under different parameters of the spacecraft mass and contact stiffness, where the contact stiffness and the equivalent mass of two spacecraft determine the natural frequency of contact dynamics. It

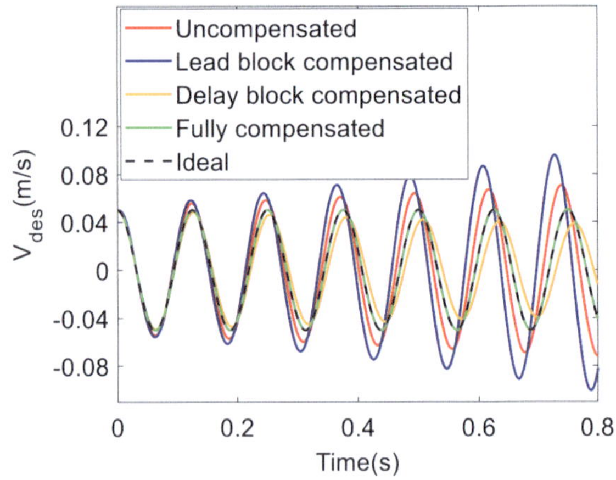

Fig. 5.18 Desired relative velocity $v_{des}(t)$ of two spacecraft simulated under parameters m_{low}: 100 kg, f_{low}: 30 Hz, ξ_{low}: 0.8, m_{spac}: 50 kg, f_{con}: 8 Hz, v_{init}: 0.05 m/s

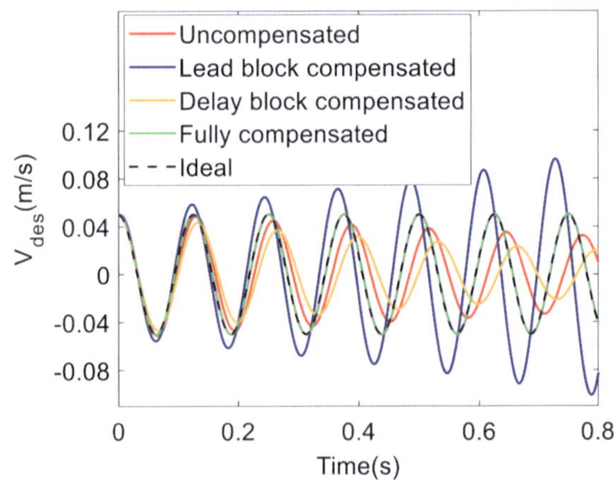

Fig. 5.19 Desired relative velocity $v_{des}(t)$ of two spacecraft simulated under parameters m_{low}: 100 kg, f_{low}: 30 Hz, ξ_{low}: 0.8, m_{spac}: 200 kg, f_{con}: 8 Hz, v_{init}: 0.05 m/s

can be seen that the divergence distortion is more serious at higher contact natural frequencies without compensation.

The above simulations illustrate that if the delay block is ignored in the compensation and only the lead block is compensated, then the system will even suffer from more serious divergence distortion because the delay block still exists. If the lead block is ignored in the compensation and only the delay block is compensated, then the system will suffer from more convergence distortion because the lead block still exists. After full compensation, the desired relative velocity of the two spacecraft is almost equal to the ideal case. The proposed distortion compensation method for the structural dynamics of the lower motion simulator can make the HIL simulation system perform a stable simulation with high accuracy under different equivalent masses of two spacecraft and different contact stiffness of contact dynamics.

Figure 5.20 shows the CoR of the uncompensated, lead-block-compensated, delay-block-compensated, and fully compensated methods under different masses

Table 5.1 CoR under different parameters of the spacecraft mass and contact stiffness

m_{low} (kg)	f_{low} (Hz)	ξ_{low}	f_{con} (Hz)	m_{spac} (kg)	k_{con} (N/mm)	v_{init} (m/s)	Ideal CoR	CoR without comp	CoR with lead block comp	CoR with delay block comp	CoR with full comp
100	30	0.8	8	50	126.33	0.05	1	1.0487	1.1043	0.9572	1.0000
100	30	0.8	8	200	505.32	0.05	1	0.9195	1.1043	0.8568	1.0000
100	30	0.8	5	50	49.35	0.05	1	1.0117	1.0241	0.9888	1.0000
100	30	0.8	5	200	197.39	0.05	1	0.9782	1.0241	0.9584	1.0000

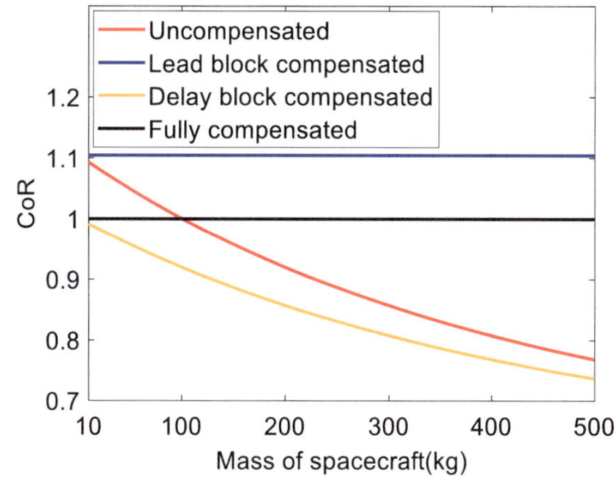

Fig. 5.20 CoR under different m_{spac} (m_{low}: 100 kg, f_{low}: 30 Hz, ξ_{low}: 0.8, f_{con}: 8 Hz, v_{init}: 0.05 m/s)

of the two spacecraft. Figure 5.21 shows the CoR of the uncompensated, lead-block-compensated, delay-block-compensated, and fully compensated methods under different contact natural frequencies. Using the proposed full compensation method, the distortion caused by the structural dynamics of the lower motion simulator can be eliminated under different spacecraft masses and different contact natural frequencies.

5.6.2 Verifications by Experiments

A HIL simulation system for space docking is developed, as shown in Fig. 5.22a. The proposed compensation method is verified through experiments. As shown in Fig. 5.22b, an undamped elastic contact experimental device is used to test the proposed compensation method. Currently, we can only design a one-dimensional undamped elastic contact experimental device in Fig. 5.22b to evaluate the accuracy of the HIL simulation. The undamped elastic contact experimental device consists

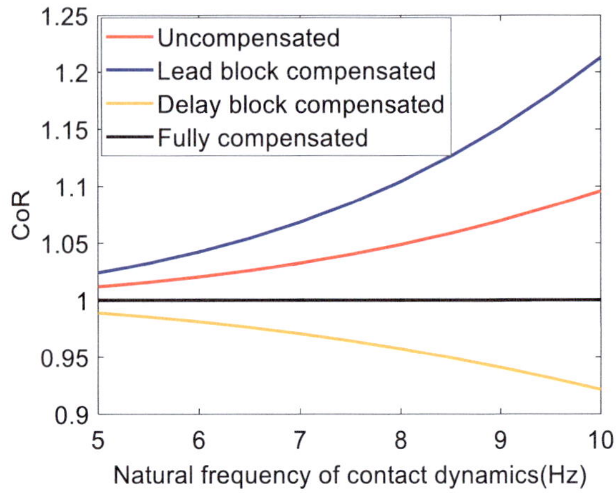

Fig. 5.21 CoR at different f_{con} (m_{low}: 100 kg, f_{low}: 30 Hz, ξ_{low}: 0.8, m_{spac}: 50 kg, v_{init}: 0.05 m/s)

of an undamped elastic rod and a high-stiffness rectangular frame. The undamped elastic rod is installed on the lower motion simulator and moves with the lower motion simulator. The rectangular frame is installed on the upper platform and fixed during the docking experiment. If the rod and the frame are placed in space as the active docking mechanism and the passive docking mechanism, then the contact result is known to be a constant-amplitude oscillation system (CoR = 1). For the install configuration in Fig. 5.22b, we can implement two one-dimensional contact experiments along a vertical direction and along a horizontal direction. By rotating the contact experimental device 90° along the vertical direction, we can carry out the third one-dimensional contact experiment along another horizontal direction. Therefore, we use three one-dimensional contact experiments to evaluate the accuracy of the HIL simulation.

Figures 5.23 and 5.24 show the CoR of the uncompensated, lead-block-compensated, delay-block-compensated, and fully compensated methods at natural frequencies of contact dynamics f_{con} 4 Hz and 8 Hz. The initial docking velocity v_{init} is 0.04 m/s. For the two spacecraft, the equivalent mass m_{spac} is 100 kg, which is less than the equivalent mass m_{low} of the lower motion simulator. For different natural frequencies f_{con} of contact dynamics, 4 Hz or 8 Hz, different contact stiffness of the undamped elastic rod is selected. If the HIL simulation system is not compensated, then divergence distortion arises (red line, CoR: 1.1227 at 4 Hz, 1.2713 at 8 Hz). Only compensating for the lead block results in more significant divergence distortion (blue line, 1.2178 at 4 Hz, CoR: 1.5408 at 8 Hz). Only compensating for the delay block results in convergence distortion (yellow line, CoR: 0.9755 at 4 Hz, 0.9495 at 8 Hz). Using the proposed full compensation method, the HIL simulation result (green line, CoR: 0.9957 at 4 Hz, 0.9930 at 8 Hz) approximates the ideal case (CoR: 1.0000). The experiments verified that the proposed compensation method is effective. The results when the equivalent mass m_{spac} of the spacecraft is larger than the equivalent mass m_{low} of the lower motion simulator are also matched with the theoretical analysis. These results are ignored here.

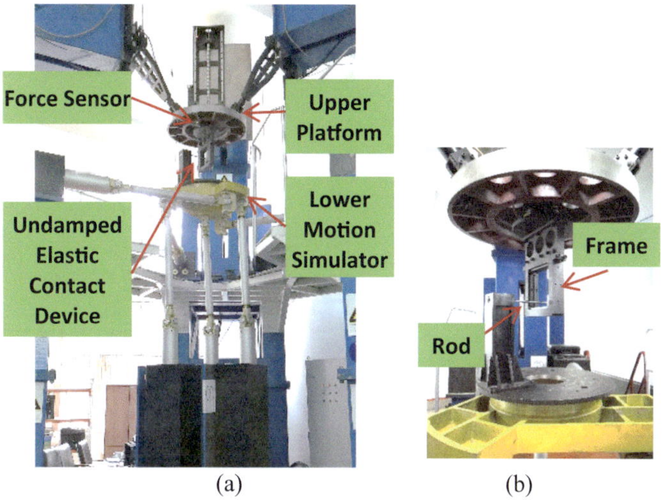

Fig. 5.22 Developed HIL simulator of space docking and undamped elastic contact experimental device

Fig. 5.23 CoR at natural frequency of contact dynamics f_{con} of 4 Hz in experiment

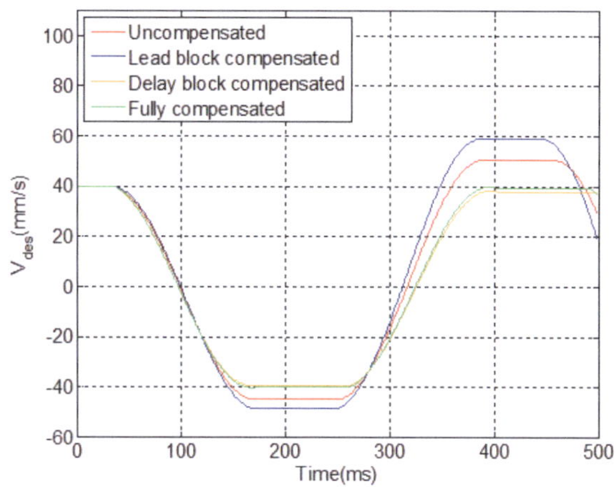

5.6.3 Applications

As shown in Fig. 5.25, the developed HIL simulator is applied to the space docking simulation of a capture mechanism in Fig. 5.26. The six-dimensional version of the proposed compensation method for the structural dynamics of the lower motion simulator is used. A 6-DOF control system is developed, where six one-dimensional distortion compensation models are designed to compensate the simulation distortion. In the experiment, the active and passive docking mechanisms approach each other

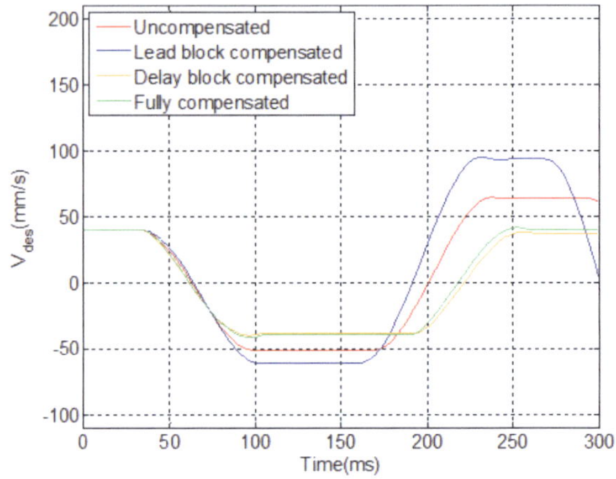

Fig. 5.24 CoR at natural frequency of contact dynamics f_{con} of 8 Hz in experiment

along the axial direction at an initial relative velocity. The active docking mechanism gradually retracts its capture claw and clasps the passive docking mechanism.

As the docking simulation continues, two docking mechanisms approach each other, and the relative motion is constrained in a limited space. In this case, reciprocating contact between the docking mechanisms occurs, and the contact natural frequency is similar in a short period. Figures 5.27 and 5.28 exhibit the three-dimensional velocity and angular velocity between two docking mechanisms during part of the reciprocating contact process at a contact natural frequency of approximately 3 Hz. The velocity rises or falls rapidly in 0.1 to 0.2 s when contact occurs and remains unchanged when the docking mechanisms move freely in the gap. The

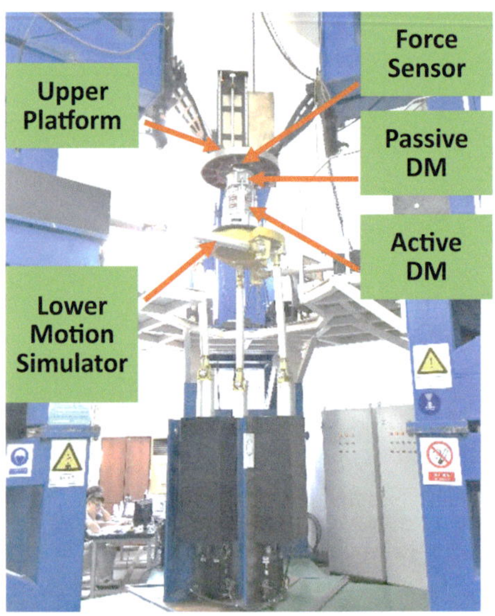

Fig. 5.25 Capture experiment on capture mechanism

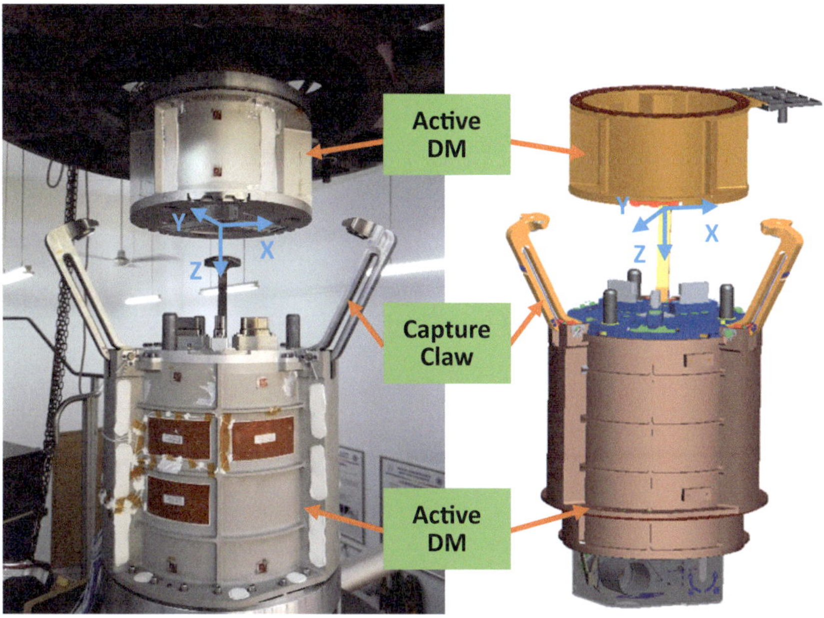

Fig. 5.26 Capture mechanism

figures show that the relative velocity between the two docking mechanisms gradually converges.

Figure 5.29 exhibits the corresponding kinetic energy of the two spacecraft. It is similar to a square waveform because of the transformation between the kinetic energy and the potential energy. The trend of the peak values of the kinetic energy reflects the increase or decrease in the system energy over time, which corresponds to

Fig. 5.27 Linear relative velocity between two docking mechanisms at contact natural frequency of approximately 3 Hz

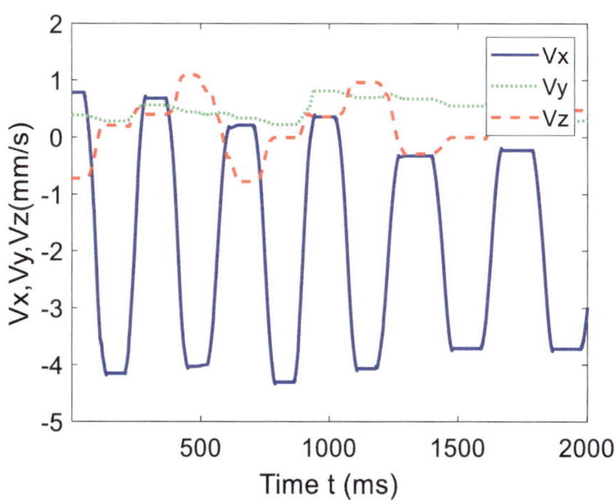

Fig. 5.28 Angular relative velocity between two docking mechanisms at contact natural frequency of approximately 3 Hz

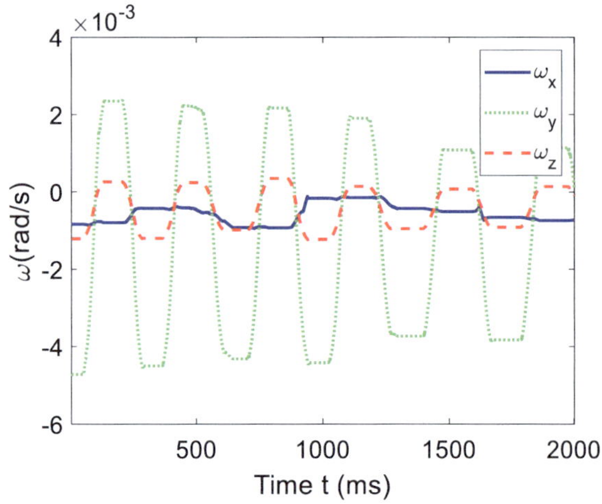

the divergence or convergence of the velocity. On the one hand, extra energy is added into the HIL simulation system, as the active mechanism drives its claws to capture the passive mechanism. On the other hand, the energy of the HIL system dissipates because of the friction and contact damping between the two docking mechanisms. Therefore, during the docking experiment, the system energy increases when elastic contact dominates the whole docking process and decreases when friction and contact damping play a dominant role. In general, the trend of the total kinetic energy appears to decrease, which means that the HIL simulation system is convergent.

The purpose of the HIL simulator is to reproduce the contact process of docking mechanisms during the spacecraft docking with high accuracy. The docking experiments can be performed on the developed HIL simulator under different configuration and docking errors between active docking mechanism and passive docking mechanisms, e.g., different mass of spacecraft, different contact stiffness, different

Fig. 5.29 Kinetic energy of docking spacecraft at contact natural frequency of approximately 3 Hz

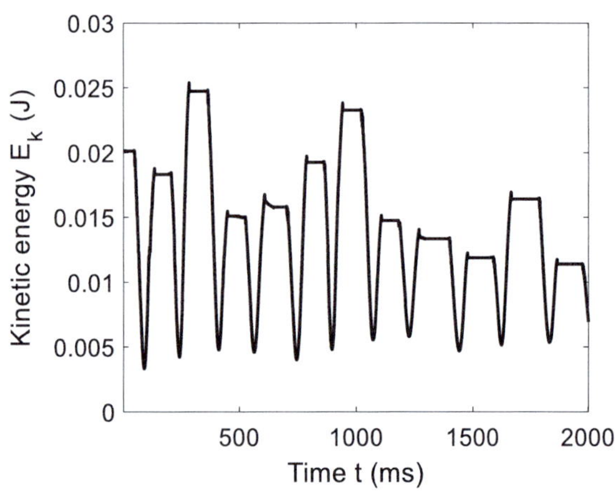

contact natural frequency, different control mode of docking mechanisms, different docking initial conditions (relative position and orientation, relative velocity and angular velocity of two spacecraft). In the future, the developed HIL simulator will have more applications.

5.7 Summary

In this chapter, how the structural dynamics of the lower motion simulator affect the accuracy of the HIL simulation is studied. It is found that due to the structural dynamics of the lower motion simulator, divergence and convergence of the HIL simulation are both possible. The conditions of the divergence and convergence of the HIL simulation due to the lower motion simulator structural dynamics are given. The distortion level of the system divergence and convergence is studied. A distortion compensation method for the HIL simulation distortion due to the structural dynamics of the lower motion simulator is proposed. An error analysis and stability analysis after the distortion compensation are given. The software emulations and experiments are used to verify the analysis and the distortion compensation method. Application experiments on real docking mechanisms are also conducted.

References

Abiko, S., Satake, Y., Jiang, X., Tsujita, T., & Uchiyama, M. (2014). Time delay compensation based on coefficient of restitution for collision hybrid motion simulator. Advanced Robotics, 28(17), 1177–1188.

Ananthakrishnan, S., Teders, R., & Alder, K. (1996). Role of estimation in real-time contact dynamics enhancement of space station engineering facility. IEEE Robotics and Automation Magazine, 3(3), 20–28.

Gao, F., Qi, C. K., Ren, A. Y., Zhao, X. C., et al. (2016). Hardware-in-the-loop simulation for the contact dynamic process of flying objects in space. Science China Technological Science, 59(8), 1167–1175.

Golubev, Y. F., Yaskevich, A. V. (2020). Hybrid simulation of spacecraft berthing. Journal of Computer and Systems Sciences International, 59(4), 609–621.

Hu, Y., Gao, F., Cao, R., & Zhao, X. C. (2017). The method of modeling of a novel 6-dof parallel manipulator as a generalized virtual road vehicle for on-board equipment test. Proceedings of the ASME 2017 International Design Engineering Technical Conferences and Computers and Information in Engineering Conference, 6–9 August, 2017, Cleveland, Ohio, USA.

Insam, C., Peiris, L. D. H., & Rixen, D. J. (2021). Normalized passivity control for hardware-in-the-loop with contact. International Journal of Dynamics and Control, 9(4), 1471–1477.

Motaghedi, P., & Stamm, S. (2005). 6 DOF testing of the orbital express capture system. Proceedings of SPIE, 5799, 66-81.

Ogata, K. (2009). Modern control engineering—Fifth edition, Prentice Hall.

Osaki, K., Konno, A., & Uchiyama, M. (2010). Time delay compensation for a hybrid simulator. Advanced Robotics, 24(8–9), 1081–1098.

Qi, C. K., Gao, F., Zhao, X. C., Wang, Q., & Sun, Q. (2018). Distortion compensation for a robotic hardware-in-the-loop contact simulator. IEEE Transactions on Control Systems Technology, 26(4), 1170–1179.

Qi, C. K., Wang, W. L., Li, D. J., Hu, Y., & Gao, F. (2021). Convergence compensation for space contact semiphysical simulator based on mechanical structural dynamics. Journal of Aerospace Engineering, 34(5), 04021066.

Qi, C. K., Li, D. J., Ma, W., Wei, Q. Q., Zhang, W. M., Wang, W. L., Hu, Y., & Gao, F. (2022). Distributed delay compensation for a hybrid simulation system of space manipulator capture. IEEE/ASME Transactions on Mechatronics, 27(4), 2367–2378.

Qi, C. K., Li, D. J., Hu, Y., Gao, F., Wang, W. L., & Ma, W. (2023). Stability analysis and distortion compensation of robot structural dynamics for a hybrid simulator. Journal of Mechanical Design, 145(7): 073302.

Chapter 6
Compensation of Structural Dynamics of Upper Platform

6.1 Introduction

A HIL simulator for space docking consists of a motion simulator, real docking mechanisms, a contact force and torque measurement system, and a numerical simulation system for the dynamics of the spacecraft. The motion simulator includes a lower platform (i.e., a lower motion simulator) and an upper platform (i.e., an upper motion simulator that can be a fixed or movable platform). It is known that time delays in the HIL simulation system of space contact lead to a simulation divergence distortion and instability problem. Some time delay compensation methods can prevent the simulation instability and divergence distortion.

The simulation convergence distortion can also occur due to dissipative factors. The simulation convergence means that the simulation error (i.e., the simulation result minus the ideal result) is negative and becomes larger and larger with time. The HIL simulation has a convergence problem due to mechanical structural dynamics of the upper platform.

In this chapter, the HIL simulation convergence problem caused by mechanical structural dynamics of the upper platform is studied (Qi et al. 2021). The phase analysis shows that the mechanical structural dynamics of the upper platform is a phase lead loop. That is why the HIL simulation convergence error increases with time. To compensate the convergence error, a mechanical structural dynamics based compensation method is investigated. The phase analysis and stability analysis are given. From simulations and experiments, it is shown that this compensation method can significantly reduce the simulation error and improve the accuracy. This compensation method can be used for the HIL simulators of space contact, where it is difficult to make the simulator perfect such that the mechanical structural dynamics of the upper platform have effects on the simulation fidelity.

In the following, the HIL simulation system is presented in Sect. 6.2. Section 6.3 discusses the HIL simulation convergence problem. A HIL simulation convergence

C. Qi et al., *Control of Hardware-In-The-Loop Space Contact Simulation Robot*,
https://doi.org/10.1007/978-981-95-1172-3_6

compensation approach is given in Sect. 6.4. Section 6.5 presents verifications and applications. The last section is the summary.

6.2 Modeling of HIL Simulation System for Space Contact

As shown in Fig. 6.1, the loop models of the HIL simulation system are developed for the analysis.

For the upper platform, its mechanical structural dynamics lead to a mechanical structural deformation $d_{stru}(t)$ under the docking force $-f_{act}(t)$, where $f_{act}(t)$ denotes the docking force acting on the active capture mechanism. The mechanical structural deformation around a working point is small, there it can be approximated by a linear dynamic system

$$m_{stru}\frac{d(d_{stru}^2(t))}{dt^2}+b_{stru}\frac{d(d_{stru}(t))}{dt}+k_{stru}d_{stru}(t)=-f_{act}(t) \tag{6.1}$$

where m_{stru}, b_{stru} and k_{stru} are the parameters of upper simulator mass, mechanical structural damping and mechanical structural stiffness. The natural frequency is $\omega_{stru}=\sqrt{k_{stru}/m_{stru}}$ (rad/s) or $f_{stru}=\omega_{stru}/2\pi$ (Hz), and the damping ratio is

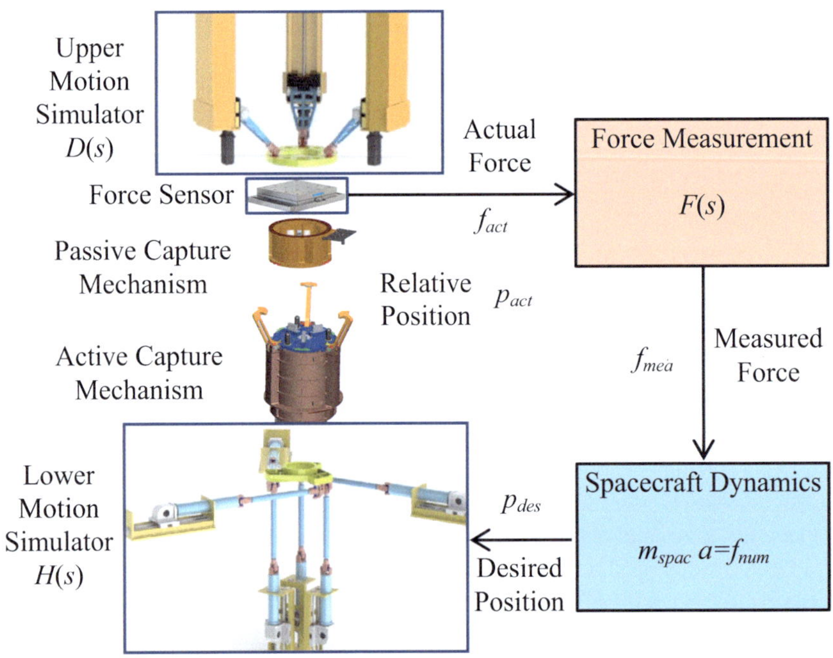

Fig. 6.1 Loop models of HIL simulator

$\xi_{stru} = b_{stru}/(2\sqrt{m_{stru}k_{stru}})$. This model can be identified using an identification method of a linear dynamical system. The identification experiment can be conducted by measuring the end-effector motion of the upper platform after a known external force is acted.

For the lower motion simulator, the desired motion $p_{des}(t)$ equals the relative motion $p_{num}(t)$ of the spacecraft that is obtained from the simulation model of spacecraft dynamics, but the actual motion is $p_{sim}(t)$ due to the dynamic response capability. Dynamic response delay of the lower motion simulator can be compensated by some compensators (Ananthakrishnan et al. 1996; Qi et al. 2017). Because this study focuses on the convergence problem from mechanical structural dynamics, it is assumed that the dynamic response delay of the lower motion simulator has been compensated very well. That is

$$H(s) = I \text{ and } p_{sim}(t) = p_{des}(t). \tag{6.2}$$

For the capture mechanisms, the elastic contact is considered for simplicity. Therefore, the actual docking force $f_{act}(t)$ can be written as

$$f_{act}(t) = -k_{con}p_{act}(t) \tag{6.3}$$

where k_{con} is the contact stiffness. The actual relative motion $p_{act}(t)$ of the capture mechanisms is

$$p_{act}(t) = p_{sim}(t) - d_{stru}(t) \tag{6.4}$$

where $p_{sim}(t)$ is the actual motion of the lower motion simulator, $d_{stru}(t)$ is the mechanical structural deformation of the upper platform. Of course, the model (6.3) can also include a damping term, then the following equations need some revisions.

For the force sensor, the actual force $f_{act}(t)$ is delayed to the measured force $f_{mea}(t)$ due to the sensor response and signal processing. This study focuses on the convergence problem from mechanical structural dynamics, therefore it is assumed that the force measurement delay has been compensated well. That is

$$F(s) = I \text{ and } f_{mea}(t) = f_{act}(t). \tag{6.5}$$

Though two spacecraft have two dynamic models, it can be proved that their relative motion to the docking force can be expressed by one dynamic model with equivalent mass of two spacecraft. Therefore, we have

$$m_{spac}a = f_{num}(t) \tag{6.6}$$

where m_{spac} is the equivalent mass, $f_{num}(t)$ is the input force of the simulation model of the equivalent spacecraft. $p_{num}(t)$ is the relative motion of two spacecraft, which is obtained by integrating the acceleration a based on the initial conditions (e.g., initial velocity and position). Note that $f_{num}(t)$ is equal to $f_{mea}(t)$ when there is no compensation. The natural angle frequency of the contact is $\omega_{con} = \sqrt{k_{con}/m_{spac}}$.

6.3 HIL Simulation Convergence

6.3.1 Convergence Phenomenon

In the following, we will show that the mechanical structural dynamics of the upper platform will result in the simulation convergence. It means that the exit velocity after the contact is smaller than the entry velocity before the contact, and therefore the CoR is less than one.

In the simulation, the system parameters are set as: docking mechanism stiffness 100N/mm, docking natural frequency 5Hz, entry velocity 100mm/s and mechanical structural mass 100kg of the upper platform. These parameters are selected only for an example, while other parameter settings will obtain similar results which does not change the study conclusions. The sampling time is 1ms.

Figure 6.2 presents the simulated actual relative velocity of two spacecraft with natural frequency 30Hz, 25Hz and 20Hz and damping ratio 0.2 of mechanical structural dynamics of the upper platform. The corresponding CoR is shown in Fig. 6.3. The ideal relative velocity of two spacecraft for the elastic contact process is a simple harmonic motion. If the natural frequency of mechanical structure is smaller, the simulation has a larger convergence.

Figure 6.4 presents the actual velocity with damping ratio 0.1, 0.4 and 0.8 and natural frequency 20Hz of the mechanical structure of the upper platform. The corresponding CoR is shown in Fig. 6.5. If the damping ratio of mechanical structure is larger, the simulation has a greater convergence.

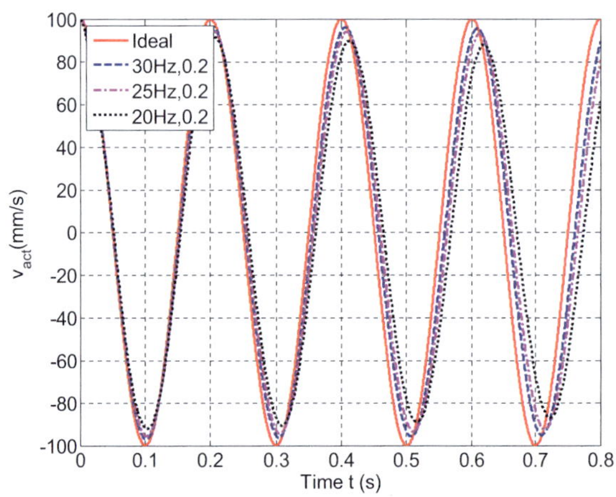

Fig. 6.2 Actual velocity with natural frequency 30Hz, 25Hz, 20Hz and damping ratio 0.2 of mechanical structure of the upper platform

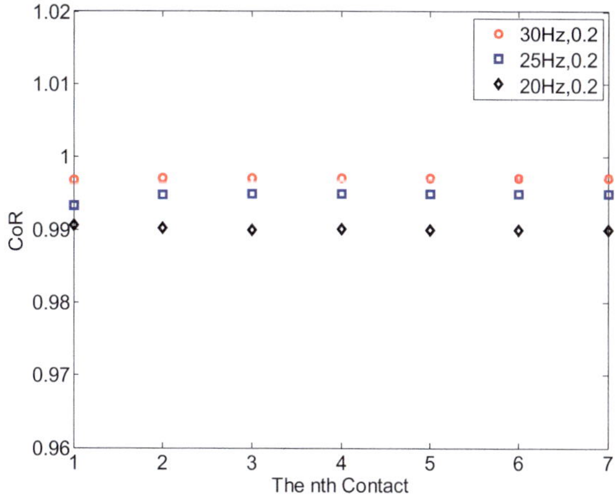

Fig. 6.3 CoR with natural frequency 30Hz, 25Hz, 20Hz and damping ratio 0.2 of mechanical structure of the upper platform

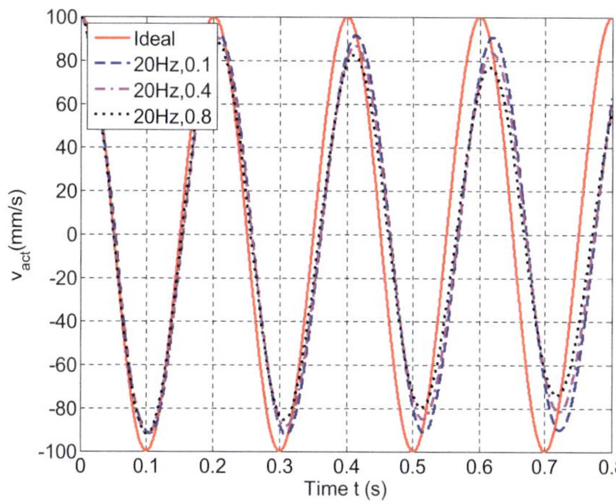

Fig. 6.4 Actual velocity with damping ratio 0.1, 0.4, 0.8 and natural frequency 20Hz of mechanical structure of the upper platform

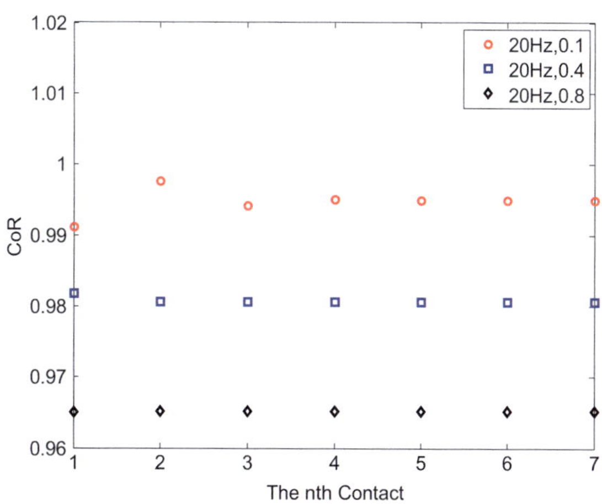

Fig. 6.5 CoR with damping ratio 0.1, 0.4, 0.8 and natural frequency 20Hz of mechanical structure of the upper platform

6.3.2 Phase Analysis of Simulation System

The simulation convergence can be explained by the phase analysis. From the models 6.1, 6.3 and 6.4, we have

$$P_{act}(s) = P_{sim}(s) - \frac{k_{con}P_{act}(s)}{m_{stru}s^2 + b_{stru}s + k_{stru}} \tag{6.7}$$

Thus

$$\frac{P_{act}(s)}{P_{sim}(s)} = \frac{m_{stru}s^2 + b_{stru}s + k_{stru}}{m_{stru}s^2 + b_{stru}s + k_{stru} + k_{con}} \tag{6.8}$$

From 6.3 and 6.8, we have

$$\frac{F_{act}(s)}{P_{sim}(s)} = \frac{-k_{con}(m_{stru}s^2 + b_{stru}s + k_{stru})}{m_{stru}s^2 + b_{stru}s + k_{stru} + k_{con}} \tag{6.9}$$

Considering $F_{mea}(s) = F_{act}(s)$ and $P_{des}(s) = P_{sim}(s)$, the transfer function model from the desired motion to the measured force is

$$\frac{F_{mea}(s)}{P_{des}(s)} = \frac{-k_{con}(m_{stru}s^2 + b_{stru}s + k_{stru})}{m_{stru}s^2 + b_{stru}s + k_{stru} + k_{con}} \tag{6.10}$$

The phase difference from the desired motion to the measured force is

$$\angle\frac{F_{mea}(s)}{P_{des}(s)} = \tan^{-1}\frac{b_{stru}\omega}{k_{stru} - m_{stru}\omega^2} - \tan^{-1}\frac{b_{stru}\omega}{k_{stru} + k_{con} - m_{stru}\omega^2} \tag{6.11}$$

When the mechanical structural damping of the upper platform is zero, there is no phase difference from the desired motion to measured force. However, in practice, the mechanical structural damping of the upper platform is not zero, i.e., $b_{stru} > 0$. In the simulator design, the natural frequency of mechanical structure of the upper platform can be made larger than the contact natural frequency ω_{con}, i.e., $\omega_{con} < \omega_{stru}$, where $\omega_{con} = \sqrt{k_{con}/m_{spac}}, \omega_{stru} = \sqrt{k_{stru}/m_{stru}}$. This condition can be satisfied when $k_{stru} > k_{con}$ and $m_{stru} < m_{spac}$, which is a common case in practice. Under the conditions of $b_{stru} > 0$ and $\omega_{con} < \omega_{stru}$, $\angle\frac{F_{mea}(s)}{P_{des}(s)}$ is a positive value. It means that $\angle\frac{F_{mea}(s)}{P_{des}(s)}$ is a phase lead loop. That is why the HIL simulation has a convergence error due to the mechanical structural dynamics of the upper platform. This phase analysis is matched with the simulation convergence results.

6.4 HIL Simulation Convergence Compensation

The mechanical structural deformation of the upper platform leads to a docking force error. The force error then results in a desired relative motion error of spacecraft. The system loops of the HIL simulator is closed, therefore the simulation error continuously accumulates and then the system is more convergent with time. To avoid the simulation convergence error, the force error should be compensated.

In this study, to compensate the convergence error a mechanical structural dynamics model based convergence compensation method is investigated. The basic idea is that if the measured force can be compensated to approximate the ideal docking force in space, the HIL simulation will achieve high-fidelity results. The force that needs to be compensated is calculated at each sampling time in real-time, by estimating the mechanical structural deformation of the upper platform and identifying the contact stiffness of the docking mechanisms. Figure 6.6 gives the HIL simulation convergence compensation framework.

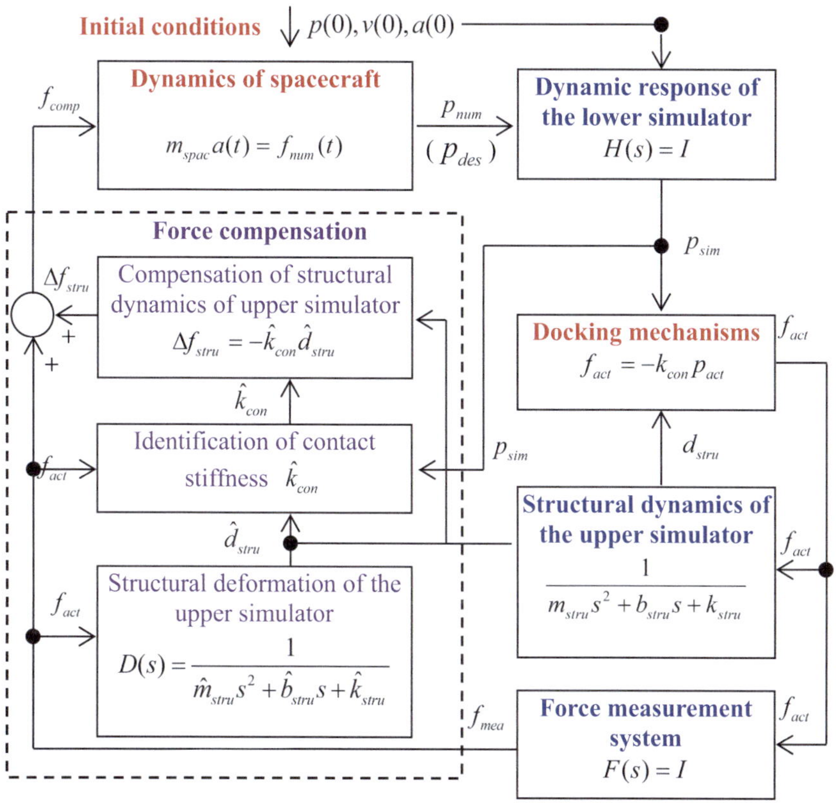

Fig. 6.6 HIL simulation convergence compensation

6.4.1 Mechanical Structural Deformation of Upper Platform

The model of mechanical structural dynamics of the upper platform can be derived from the simulator design model and calibrated from experiments before the HIL simulation. In this study, a second-order model is used as a compensation model of mechanical structural dynamics of the upper platform

$$\hat{D}_{stru}(s) = -\hat{D}(s)F_{act}(s) = -\frac{1}{\hat{m}_{stru}s^2 + \hat{b}_{stru}s + \hat{k}_{stru}}F_{act}(s) \tag{6.12}$$

where $\hat{m}_{stru}$, $\hat{b}_{stru}$ and $\hat{k}_{stru}$ are the mechanical structural mass, damping and stiffness of the compensation model respectively. The natural frequency of the compensation model is denoted as $\hat{f}_{stru} = \sqrt{\hat{k}_{stru}/\hat{m}_{stru}}/2\pi$ (Hz). Here this typical model is selected only for the simulation study. For the real mechanical structure of the upper platform, an accurate enough model obtained from the theoretical derivation and experimental calibration is necessary. Using the compensation model, the mechanical structural deformation $\hat{d}_{stru}(t)$ can be calculated from the measured docking force $f_{mea}(t)$.

6.4.2 Compensation of Mechanical Structural Dynamics of Upper Platform

In this study, to compensate the simulation convergence error a mechanical structural dynamics model based convergence compensation method is investigated. The motion error of the simulator is

$$d_{stru}(t) = p_{des}(t) - p_{act}(t) \tag{6.13}$$

The motion error will result in a docking force error $\Delta f_{stru}(t)$. The physical properties of the capture mechanisms give a description between the docking force and relative motion

$$f_{act}(t) = z_{con}(p_{act}(t)) \tag{6.14}$$

The physical properties of the capture mechanisms are assumed to be unchanged at each sampling time (i.e., 1ms). By identifying physical properties of capture mechanisms in real-time, the compensation force can be obtained. Using the Taylor expansion, the compensation force for a motion error can be expressed as

$$\Delta f_{stru}(t) = \left.\frac{\partial f_{act}}{\partial p_{act}}\right|_{p_{act}(t)} d_{stru}(t) = -k_{con}(t)d_{stru}(t) \tag{6.15}$$

For the undamped elastic contact, $k_{con}(t)$ is the contact stiffness of two docking mechanisms. For a damped contact, $k_{con}(t)$ consists of contact stiffness and damping of two docking mechanisms.

Finally, the compensated docking force is achieved as follows

$$f_{comp}(t) = f_{mea}(t) + \Delta f_{stru}(t) \tag{6.16}$$

From 6.5, 6.13, 6.15 and 6.16, the compensated docking force can also be rewritten as

$$f_{comp}(t) = -k_{con}p_{act} - k_{con}(t)d_{stru}(t) = -k_{con}p_{des}(t) = f_{ideal}(t) \tag{6.17}$$

It means that the docking force after the compensation is very similar to the ideal docking force. This is why the HIL simulation system will approximate the contact system in space.

6.4.3 *Identification of Contact Stiffness of Docking Mechanisms*

Now we consider the undamped elastic contact for simplicity but the approach is applicable to the damped cases. The actual value of relative position of capture mechanisms can be obtained from the output position $p_{sim}(t)$ of the lower motion simulator and the mechanical structural deformation $\hat{d}_{stru}(t)$ of the upper platform

$$\hat{p}_{act}(t) = p_{sim}(t) - \hat{d}_{stru}(t) \tag{6.18}$$

Using Eq. 6.15 we have

$$\Delta f_{act}(t) = -\Delta p_{act}(t)k_{con}(t) \tag{6.19}$$

When several estimated actual motion $\hat{p}_{act}(t), \hat{p}_{act}(t-1), \ldots, \hat{p}_{act}(t-n)$ and actual docking force $f_{act}(t), f_{act}(t-1), \ldots, f_{act}(t-n)$ are available, their incremental values can be constructed as follows

$$\begin{aligned}\Delta\hat{p}_i^t &= -\Delta\hat{p}_{act}(t-n+i)\\ &= \hat{p}_{act}(t-n+i-1)\\ &\quad - \hat{p}_{act}(t-n+i), i = 1, \ldots, n\end{aligned} \tag{6.20}$$

$$\begin{aligned}\Delta f_i^t &= \Delta f_{act}(t-n+i)\\ &= f_{act}(t-n+i)\\ &\quad - f_{act}(t-n+i-1), i = 1, \ldots, n\end{aligned} \tag{6.21}$$

Then the stiffness parameter $k_{con}(t)$ is estimated from $\{\Delta\hat{p}_i^t, \Delta\hat{f}_i^t\}_{i=1}^n$ using recursive least squares methods in real-time as follows

$$k_i^t = k_{i-1}^t + L_i^t(\Delta f_i^t - \Delta\hat{p}_i^t \hat{k}_{i-1}^t), i = 1, ..., n \tag{6.22}$$

$$L_i^t = \frac{R_{i-1}^t \Delta p_i^t}{1 + \Delta p_i^t R_{i-1}^t \Delta p_i^t} \tag{6.23}$$

$$R_i^t = R_{i-1}^t - \frac{R_{i-1}^t \Delta p_i^t \Delta p_i^t R_{i-1}^t}{1 + \Delta p_i^t R_{i-1}^t \Delta p_i^t} \tag{6.24}$$

where k_i^t is the stiffness value at the ith step, L_i^t is the gain, R_i^t is positive and denotes the inverse covariance and $\hat{k}_{con}(t) = k_n^t$.

6.4.4 Stability Analysis of HIL Simulator

The stability analysis of the HIL simulation closed-loop system is provided in the following. Here the compensation model of mechanical structural dynamics is assumed to be accurate, i.e., $\hat{m}_{stru} = m_{stru}$, $\hat{b}_{stru} = b_{stru}$ and $\hat{k}_{stru} = k_{stru}$. From Eqs. 6.1–6.3, 6.12, 6.15 and 6.16, the closed-loop characteristic equation can be derived as follows

$$\begin{aligned} \chi(s) = {} & m_{spac} m_{stru} s^4 + m_{spac} b_{stru} s^3 + (m_{spac} k_{stru} + m_{spac} k_{con} \\ & + m_{stru} k_{con}) s^2 + k_{con} b_{stru} s + k_{con}(k_{stru} + \hat{k}_{con}) \end{aligned} \tag{6.25}$$

The necessary and sufficient conditions of the closed-loop stability can be obtained as follows

$$m_{spac} > 0, k_{con} > 0, m_{stru} > 0, b_{stru} > 0, k_{stru} > 0, k_{con} \geq \hat{k}_{con} > 0 \tag{6.26}$$

from Routh-Hurwitz rules. It means that when the estimated stiffness is equal to or less than the real contact stiffness, the HIL simulator is stable.

From the identification algorithm 6.22–6.24, the following equation can be derived

$$k_{con} - k_i^t = \left(\frac{I}{1 + \Delta p_i^t R_{i-1}^t \Delta p_i^t}\right)(k_{con} - k_{i-1}^t) \tag{6.27}$$

Considering $\Delta p_i^t R_{i-1}^t \Delta p_i^t \geq 0$, we have

$$0 < \frac{1}{1 + \Delta p_i^t R_{i-1}^t \Delta p_i^t} \leq 1 \tag{6.28}$$

It means that $\Delta k_{con} = k_{con} - k_i^t$ tends to zero monotonously because $k_{con} - k_i^t$ has the same sign as $k_{con} - k_{i-1}^t$ and $k_{con} - k_i^t \le k_{con} - k_{i-1}^t$. Thus the condition $k_{con} \ge \hat{k}_{con} > 0$ can be guaranteed in the stiffness identification if the initial stiffness value is less than the real contact stiffness. The minimum stiffness value can be found from the offline stiffness identification experiments before the contact experiments. When the compensation model of mechanical structural dynamics is inaccurate, the analytical stability conditions are difficult to obtain and numerical methods can be used.

6.5 Verifications and Applications

6.5.1 Verifications by Software

The simulation verifications are performed on undamped elastic contacts using computer software. As an example, one-dimensional results at the x-axis are given because other dimensions have similar results. In the simulation, the parameter setting is: docking mechanism stiffness 100N/mm, docking natural frequency 5Hz, entry velocity 100mm/s, mechanical structural natural frequency 20Hz, mechanical structural damping ratio 0.8, and mechanical structural mass 100kg.

Figure 6.7 shows the actual velocity with the compensation of mechanical structural stiffness, with the compensation of mechanical structural stiffness and damping, and without the compensation. The ideal velocity is a simple cosine motion. Without any compensation, the HIL simulation result is the worst one. With the compensation of mechanical structural stiffness, the HIL simulation achieves a more accurate contact natural frequency and the amplitude convergence error still exists. With the compensation of the mechanical structural stiffness and damping, the HIL simulation is satisfactory with an accurate contact natural frequency and a very small convergence error. The CoR in Fig. 6.8 shows the result with the compensation of mechanical structural stiffness and damping, the simulation is very close to the ideal result because its CoR is close to 1. With the compensation of only mechanical structural stiffness or without the compensation, the simulation has a significant convergence error since their CoR is about 0.965. Figure 6.9 shows the docking force with the compensation of mechanical structural stiffness and with the compensation of mechanical structural stiffness and damping. With the compensation of mechanical structural stiffness and damping, the docking force is very similar to an ideal docking force. These results show that the convergence compensation approach can work well.

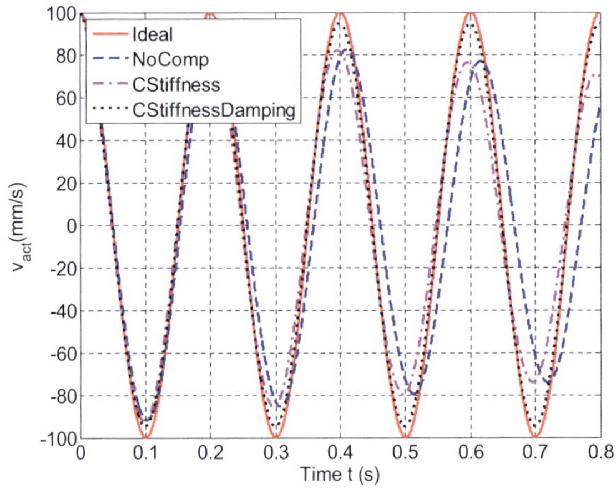

Fig. 6.7 Velocity without compensation, with compensation of mechanical structural stiffness and with compensation of mechanical structural stiffness and damping

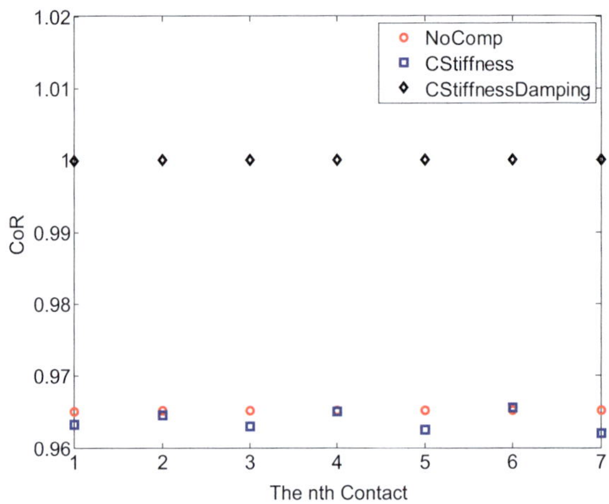

Fig. 6.8 CoR without compensation, with compensation of mechanical structural stiffness and with compensation of mechanical structural stiffness and damping

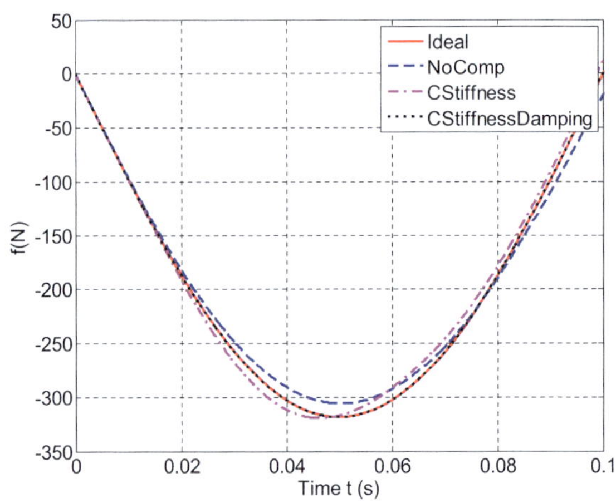

Fig. 6.9 Docking force with compensation of mechanical structural stiffness and with compensation of mechanical structural stiffness and damping

6.5.2 Verifications by Experiments

Experiments on an undamped elastic rod as shown in Fig. 6.10 are performed to test the simulation accuracy of the developed HIL simulator. One-dimensional contact results at the x-axis are given because other dimensions have similar results. Because the stiffness of the undamped elastic rod is known, the ideal docking results for a given spacecraft mass and initial entry velocity are known. By comparing the actual experimental results on the HIL simulator with the ideal docking results, the accuracy of the HIL simulator can be evaluated. That is why an undamped elastic rod is used here. A set of typical experimental parameters are selected as an example: contact stiffness of docking mechanism 105N/mm, contact natural frequency 3Hz and initial entry velocity 50mm/s.

Figure 6.11 plots the simulated relative velocity of two spacecraft without the compensation, while Fig. 6.12 gives the simulated relative velocity of two spacecraft with the compensation of mechanical structural stiffness and damping. Figure 6.13 shows the docking force without the compensation, while Fig. 6.14 provides the docking force with the compensation of mechanical structural stiffness and damping. Without the compensation, the simulation has a relatively significant error (too convergence). After the compensation of mechanical structural stiffness and damping of upper platform, the simulation error is reduced to a satisfactory level (small convergence).

From Fig. 6.15, we can see that the CoR without any compensation and with the compensation of mechanical structural stiffness and damping is about 0.982 and 0.992 respectively. The simulation convergence error is reduced with the compensation of mechanical structural stiffness and damping. Only a little convergence error about -1% is remained, which is expected to keep the system stability.

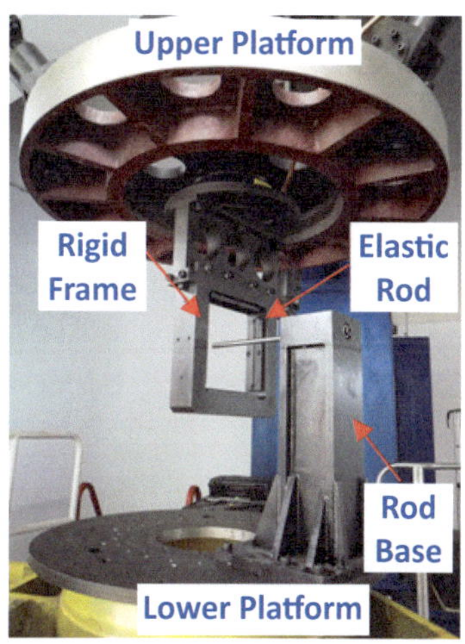

Fig. 6.10 Contact experiment device

Fig. 6.11 Velocity without compensation

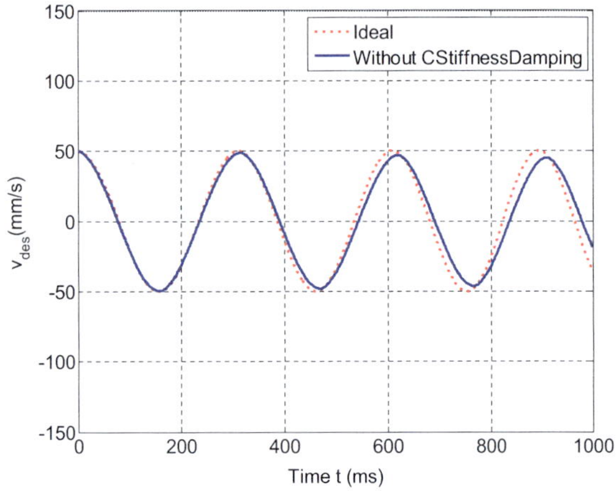

Fig. 6.12 Velocity with compensation of mechanical structural stiffness and damping

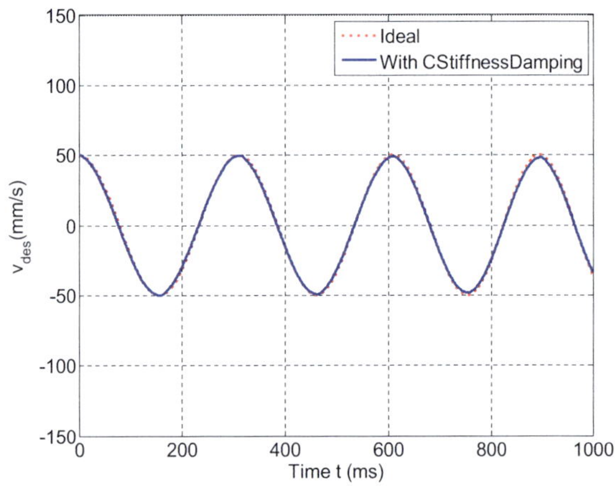

Fig. 6.13 Docking force without compensation

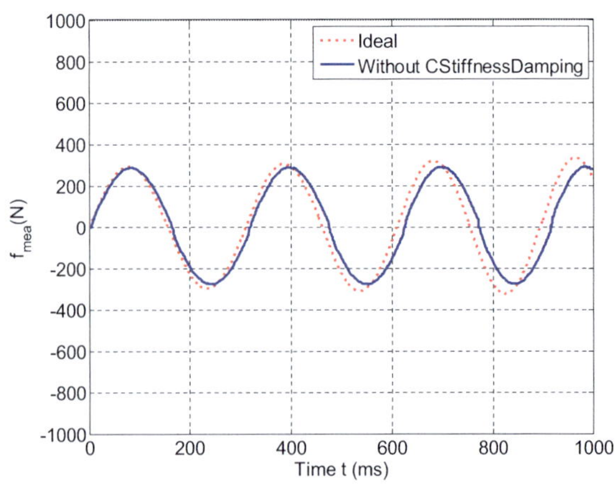

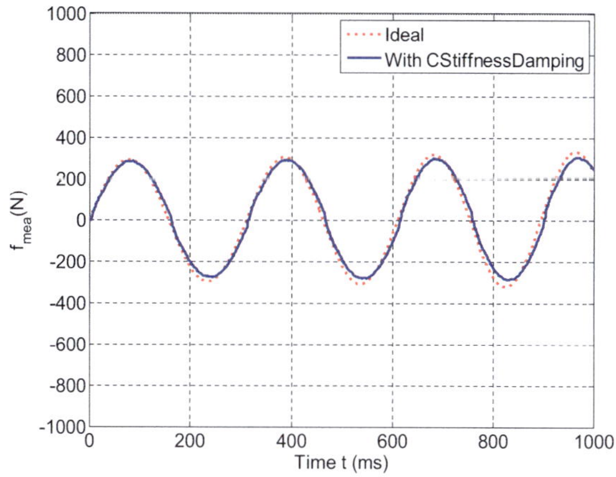

Fig. 6.14 Docking force with compensation of mechanical structural stiffness and damping

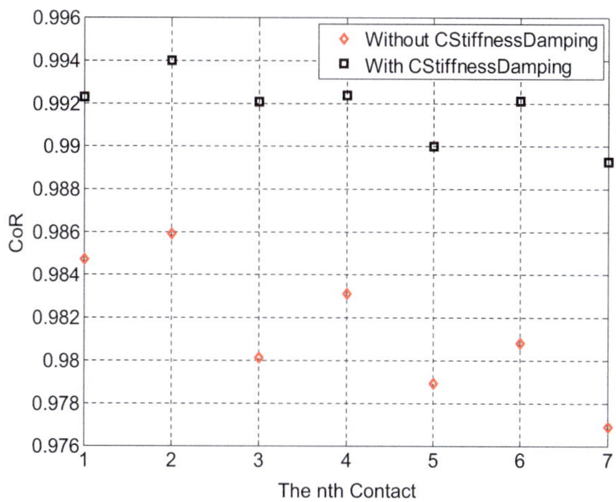

Fig. 6.15 CoR with compensation of mechanical structural stiffness and damping

Figure 6.16 plots the desired and actual positions, desired velocity, measured force, compensated force ($f_{CStiffnesDamping}$) with the compensation of mechanical structural stiffness and damping. The docking force after the compensation has a larger peak value than the measured force. The docking force after the compensation has a rightwards motion around the peak value. Therefore, the phase lead between the measured force and desired position is compensated. That is why the convergence compensation approach works. These experiments show that the mechanical structural dynamics model based convergence compensation method is effective for the HIL simulation system.

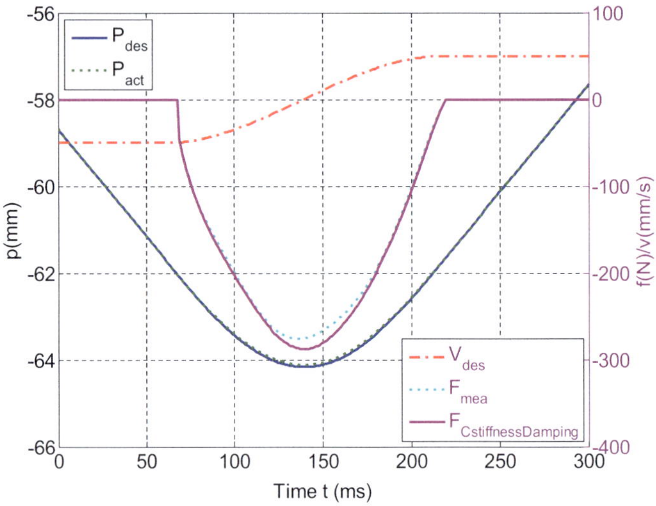

Fig. 6.16 Desired and actual position, desired velocity, measured and compensated force with compensation of mechanical structural stiffness and damping

6.5.3 Applications

The real application is a six-DOF capture experiment on a pair of three-claw capture mechanism as shown in Fig. 6.17. Given the initial capture condition including the relative position (Px = −37mm, Py = 0mm, Pz = 56mm) and zero initial relative velocity, the HIL simulator can simulate the capture process as shown in Fig. 6.18. Figure 6.19a shows relative position and velocity of docking mechanisms, and docking force curves, and Fig. 6.19b shows relative posture and angular velocity of docking mechanisms and docking torque curves. For the whole capture and docking process, the maximum docking force and torque (Fx = 61N, Fy = 107N, Fz = 96N, Mx = 13Nm, My = 18Nm, Mz = 21Nm), and the maximum motion velocity and angular velocity (Vx = 4.7mm/s, Vy = 3.9mm/s, Vz = 1.9mm/s, wx = 0.22deg/s, wy = 0.35deg/s, wz = 0.24deg/s) can be found. The capture is successful because the relative posture of the capture mechanisms approach to zero, and the initial posture offset is corrected during the capture process. The above application experiments show that developed HIL simulator with the proposed compensation approach are useful to simulate the six-DOF capture process for the real capture mechanisms.

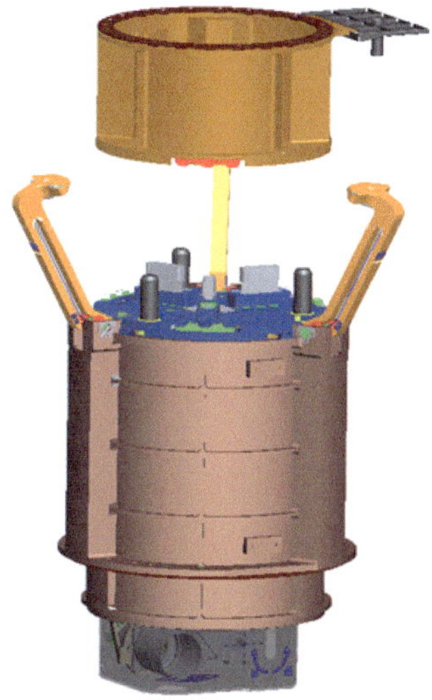

Fig. 6.17 A pair of three-claw capture mechanism

Fig. 6.18 Six-DOF capture experiment on a pair of three-claw capture mechanism

6.6 Summary

In this chapter, the simulation convergence problem due to mechanical structural dynamics of the upper platform is investigated. It is found that the phase of the measured docking force between two docking mechanisms is ahead of the desired relative position. To compensate the simulation convergence error, a mechanical

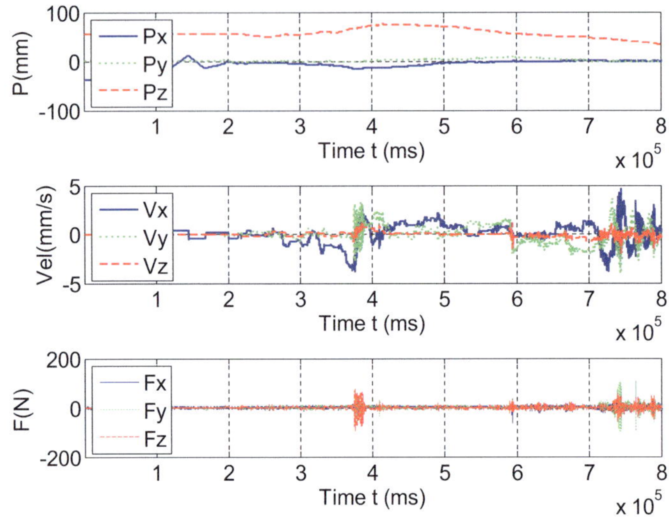

(a) Relative position, velocity, and force curves

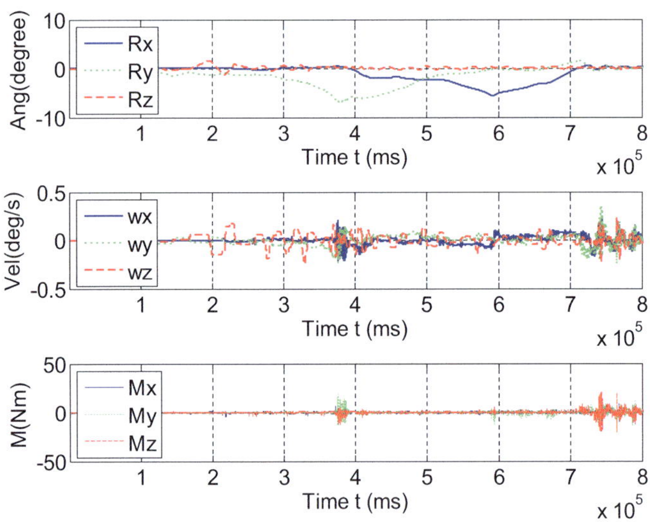

(b) Relative attitude, angular velocity and torque curves

Fig. 6.19 Six-dimensional capture experimental results on a pair of three-claw and three-slot capture mechanism structural dynamics model based convergence compensation approach is investigated. The phase analysis and stability analysis of the HIL simulation system are given. The effectiveness of the convergence compensation method is shown in simulations and experiments.

References

Ananthakrishnan, S., Teders, R., & Alder, K. (1996). Role of estimation in real-time contact dynamics enhancement of space station engineering facility. IEEE Robotics and Automation Magazine, 3(3), 20–28.

Qi, C. K., Ren, A. Y., Gao, F., Zhao, X. C., Wang, Q., & Sun, Q. (2017). Compensation of velocity divergence caused by dynamic response for hardware-in-the-loop docking simulator. IEEE/ASME Transactions on Mechatronics, 22(1), 422–432.

Qi, C. K., Wang, W. L., Li, D. J., Hu, Y., & Gao, F. (2021). Convergence compensation for space contact semiphysical simulator based on mechanical structural dynamics. Journal of Aerospace Engineering, 34(5), 04021066.

Chapter 7
Lumped Distortion Compensation Control

7.1 Introduction

For the HIL simulation of space contact, the distortion is a very important problem because it significantly affects experimental accuracy. It is more preferable to design HIL system using perfect technical parameters. However, currently it is difficult to satisfy these technical requirements using available mechanical and control systems. Therefore, the distortion problem always exists for the HIL simulation. The distortion compensation is a necessary and effective approach to improve the simulation accuracy.

The phase lag or lead of the HIL contact simulator can come from the force measurement system, the dynamic response of the lower motion simulator and the structural deformation of the upper platform. In general, the force measurement system delay is identifiable and time-invariant, thus the phase lead can be used to compensate this time delay. The dynamic response delay of the lower motion simulator is dynamic because it is related to the posture, payload and response frequency of the lower motion simulator. This dynamic delay is difficult to be known and modeled due to the complicated time-varying dynamics, thus the model-free based force compensation is expected. Because the upper platform does not move during the contact process, thus its structural dynamics can be modeled accurately and the model based force compensation can be used.

In this chapter, a HIL simulation distortion compensation approach is proposed from three aspects (Qi et al. 2018). The measured contact force depends on unknown contact stiffness of docking devices. The value of this force can be refined in real time by using estimations of contact stiffness, position errors of the lower motion simulator, structural deformations of the upper platform, and force measuring time delay. The force measurement delay is compensated by the phase lead. The dynamic response delay of the lower motion simulator is compensated by the position response error based force compensation. The structural deformation of the upper platform is compensated by the structural dynamics based force compensation. These three

C. Qi et al., *Control of Hardware-In-The-Loop Space Contact Simulation Robot*,
https://doi.org/10.1007/978-981-95-1172-3_7

compensators form a lumped force compensator because it is the measured force that is corrected. Simulations and experiments show that the proposed approach is effective to compensate the simulation distortion and improve the simulation accuracy.

This chapter is organized as follows. The system model is given in Sect. 7.2. The distortion phenomena and analysis are presented in Sect. 7.3. The distortion compensation approach is proposed in Sect. 7.4. Section 7.5 includes the verifications. A summary is given in Sect. 7.6.

7.2 HIL Simulation Modeling

Five subsystem models are established in Fig. 7.1 for the analysis of the HIL simulation system.

7.2.1 Docking Mechanisms

For simplicity, assume that the contact between docking mechanisms is dominated by the stiffness though the damping can also be considered. Moreover, a one-dimensional contact is used as an example, though the method is applicable for the six-dimensional contact. Then the actual contact force $f_{act}(t)$ can be expressed as follows

$$f_{act}(t) = -k_{con} p_{act}(t) \tag{7.1}$$

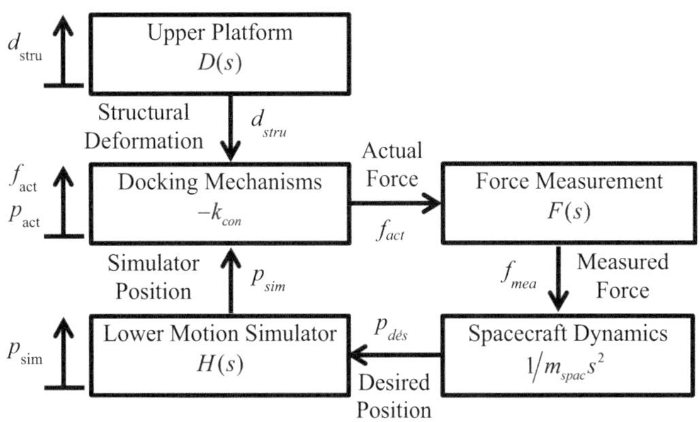

Fig. 7.1 System modeling

where k_{con} is the stiffness and t denotes the time. The actual relative position $p_{act}(t)$ of docking mechanisms is

$$p_{act}(t) = p_{sim}(t) - d_{stru}(t) \tag{7.2}$$

where $p_{sim}(t)$ is the relative position of docking mechanisms obtained directly from the lower motion simulator position without considering the structural deformation of the upper platform, $d_{stru}(t)$ is the structural deformation of the upper platform, $p_{sim} = 0$ and $d_{stru} = 0$ when two spacecraft just contact with zero force.

7.2.2 Force Measurement

Considering the force sensor, data acquisition system and noise filter, the time delay T_{fms} between the measured force $f_{mea}(t)$ and the actual force $f_{act}(t)$ can be approximated as a pure time delay

$$F_{mea}(s) = F(s)F_{act}(s) = e^{-T_{fms}s}F_{act}(s) \tag{7.3}$$

Of course, other force measurement delay can also be used with minor revision of the force measurement delay compensator.

7.2.3 Spacecraft Dynamics

The relative motion $p_{num}(t)$ of two spacecraft is simulated numerically by a spacecraft dynamics model

$$m_{spac}\ddot{p}_{num}(t) = f_{num}(t) \tag{7.4}$$

where m_{spac} is the equivalent mass of two spacecraft, $f_{num}(t)$ is the contact force for the numerical simulation. The numerically computed position $p_{num}(t)$ is also the desired position $p_{des}(t)$ of the lower motion simulator. The desired velocity $v_{des}(t)$ can be obtained from $\dot{p}_{des}(t)$. The contact natural frequency is defined as $f_{con} = 1/2\pi\sqrt{k_{con}/m_{spac}}$ (Hz).

7.2.4 Lower Motion Simulator

In this study, assume that the dynamic response of the lower motion simulator is dominated by a second-order model

$$p_{sim}(t) = H(s)p_{des}(t) = \frac{\omega_{resp}^2}{s^2 + 2\xi_{resp}\omega_{resp}s + \omega_{resp}^2} p_{des}(t) \tag{7.5}$$

where ω_{resp} is the undamped natural frequency (rad/s), ξ_{resp} is the damping ratio. The response frequency is also $f_{resp} = \omega_{resp}/2\pi$ (Hz). Note that this model is only used as an example in the analysis study, while it is not required in the distortion compensation.

The structural dynamics of the lower motion simulator is not considered here for simplicity. With minor revisions, the proposed distortion compensation method can also handle with the structural dynamics of the lower motion simulator.

7.2.5 *Upper Platform*

The structural dynamics of the upper platform will lead to the structural deformation $d_{stru}(t)$ due to the contact force $-f_{act}(t)$ acting on the upper platform. In the simulation study, the structural dynamics of the upper platform is assumed to be described by a second-order model as follows

$$D_{stru}(s) = -D(s)F_{act}(s) = -\frac{F_{act}(s)}{m_{stru}s^2 + b_{stru}s + k_{stru}} \tag{7.6}$$

where m_{stru} is the structural mass of the upper platform, b_{stru} is the structural damping of the upper platform, k_{stru} is the structural stiffness of the upper platform. $\omega_{stru} = \sqrt{k_{stru}/m_{stru}}$ is the undamped natural frequency (rad/s) of the structural dynamics, $\xi_{stru} = b_{stru}/(2\sqrt{m_{stru}k_{stru}})$ is the damping ratio of the structural dynamics. The structural natural frequency is also $f_{stru} = \omega_{stru}/2\pi$ (Hz). Note that this model is only used as an example in the analysis study, while the accurate structural dynamics model of the upper platform should be established for the real HIL simulator in the distortion compensation.

7.3 HIL Simulation Distortion

In this section, we will show that the force measurement delay and the dynamic response delay of the lower motion simulator are both phase lag factors, thus they make the HIL simulation divergent, i.e., CoR > 1. The structural dynamics of the upper platform is a phase lead factor, thus it makes the HIL simulation convergent, i.e., CoR < 1. In the simulation, the contact stiffness is 100N/mm, the contact natural frequency is 5Hz and the initial velocity is 100mm/s as an example.

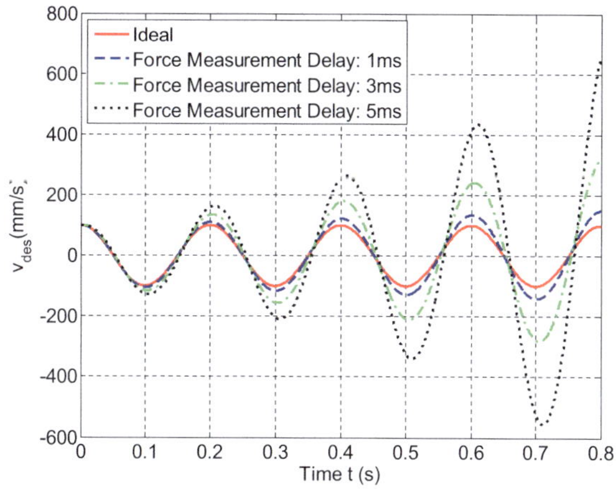

Fig. 7.2 Desired velocity without compensation at different force measurement delay

7.3.1 *Divergence Caused by Force Measurement Delay*

Assuming that the lower motion simulator and the upper platform are ideal, that is $H(s) = 1$ and $D(s) = 0$, only the force measurement delay exists in the simulation. Figure 7.2 shows the desired velocity with ideal, 1ms, 3ms and 5ms force measurement delay, where the CoR is 1.0, 1.0506, 1.1591 and 1.2776 respectively. Obviously, as the force measurement delay increases, the HIL simulation becomes more divergent.

7.3.2 *Divergence Caused by Robot Dynamic Response Delay*

Assuming that the force measurement system and the upper platform are ideal, that is $F(s) = 1$ and $D(s) = 0$, only the dynamic response of the lower motion simulator exists in the simulation. Figure 7.3 shows the desired velocity with ideal, 100Hz, 50Hz and 25Hz frequency response, where the CoR is 1.0, 1.1335, 1.2825 and 1.6260 respectively. As the frequency response of the lower motion simulator decreases, the HIL simulation becomes more divergent. Here the damping ratio of the dynamic response of the lower motion simulator is set to 0.8, while similar results are achieved for other damping ratios less than 1.

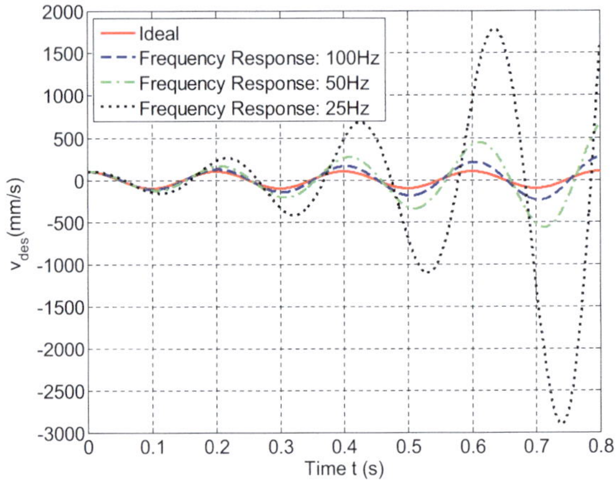

Fig. 7.3 Desired velocity without compensation at different frequency response of the lower motion simulator

7.3.3 Convergence Caused by Upper Platform Structural Dynamics

Assuming that the force measurement system and the lower motion simulator are ideal, that is $F(s) = 1$ and $H(s) = 1$, we study the effect of the structural dynamics of the upper platform on the simulation. Figure 7.4 shows the desired velocity with ideal, 30Hz, 25Hz and 20Hz natural frequency and 0.2 damping ratio of the structural dynamics of the upper platform, where the CoR is 1.0, 0.9970, 0.9946 and 0.9901 respectively. When the natural frequency of structural dynamics of the upper platform is smaller, the HIL simulation becomes more convergent. Figure 7.5 presents the desired velocity with ideal, 0.05, 0.1 and 0.2 damping ratio and 20Hz natural frequency of the structural dynamics of the upper platform, where the CoR is 1.0, 0.9968, 0.9947 and 0.9901 respectively. It can be seen that when the damping ratio of structural dynamics of the upper platform is larger, the HIL simulation becomes more convergent.

7.4 Lumped Distortion Compensation Approach

7.4.1 Lumped Distortion Compensation Framework

The distortion compensation principle is that we can compensate the measured force so that it approximates the ideal force in real-time. Then the HIL simulation result will approach the real space contact without the distortion. The lumped distortion compensation framework is presented in Fig. 7.6, where the force measurement delay is compensated by the phase lead; the upper platform is compensated by the structural

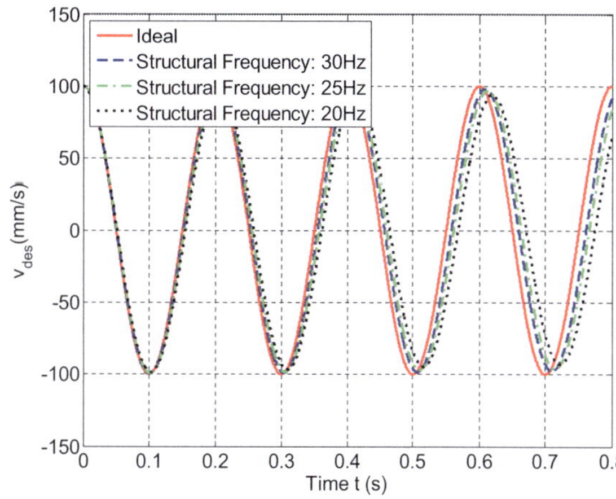

Fig. 7.4 Desired velocity without compensation at different natural frequency of structural dynamics of the upper platform

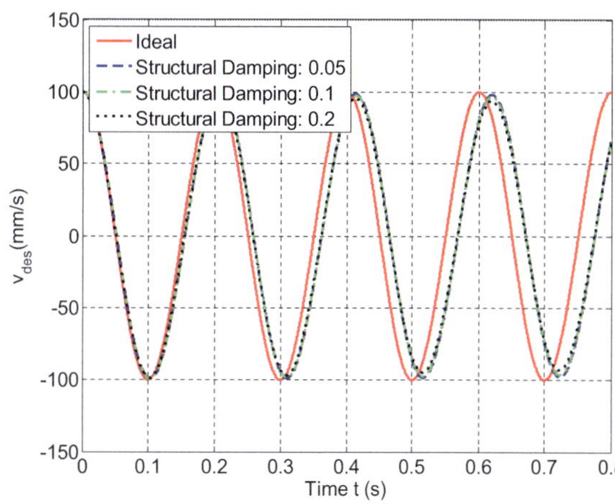

Fig. 7.5 Desired velocity without compensation at different damping ratio of structural dynamics of the upper platform

deformation based force compensation; the dynamic response delay of the lower motion simulator is compensated by the position error based force compensation.

7.4.2 Force Measurement Delay Compensation

In theory, the time lead $e^{T_{fms}s}$ is a good compensation for $e^{-T_{fms}s}$. The Taylor expansion of $e^{T_{delay}s}$ can be expressed by

$$e^{T_{delay}s} = 1 + T_{delay}s + \frac{T_{delay}^2 s^2}{2} + ... \tag{7.7}$$

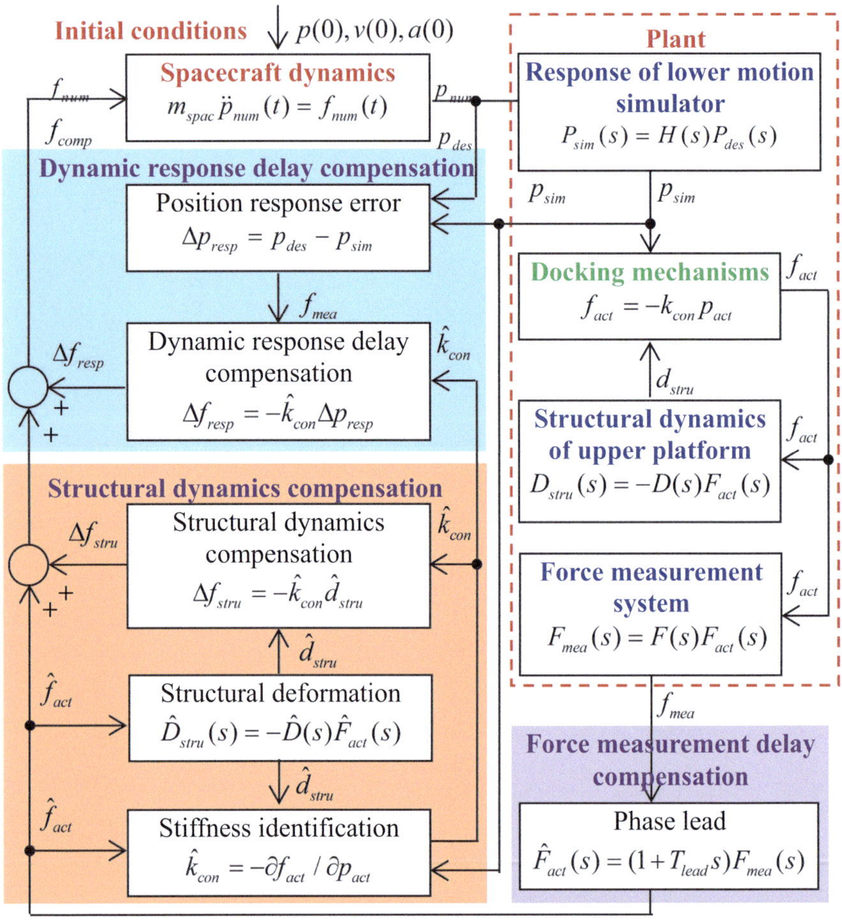

Fig. 7.6 Lumped distortion compensation framework

In the practical compensation, the high-order terms are ignored as follows

$$\hat{F}_{act}(s) = (1 + T_{lead}s)F_{mea}(s) \tag{7.8}$$

The time lead T_{lead} will be equal to T_{delay}, which can be estimated in advance from the measured force and the actual position data.

7.4.3 Upper Platform Structural Dynamics Compensation

(1) Structural Deformation of the Upper Platform

The model of the structural dynamics of the upper platform can be established or identified from experiments. In the simulation study, assume that a compensation model for the structural dynamics of the upper platform is described by a second-order model as follows

$$\hat{D}_{stru}(s) = -\hat{D}(s)\hat{F}_{act}(s) = -\frac{1}{\hat{m}_{stru}s^2 + \hat{b}_{stru}s + \hat{k}_{stru}}\hat{F}_{act}(s) \tag{7.9}$$

where $\hat{m}_{stru}$ is structural mass of the compensation model of the structural dynamics, $\hat{b}_{stru}$ is the structural damping of the compensation model of the structural dynamics, and $\hat{k}_{stru}$ is the structural stiffness of the compensation model of the structural dynamics. The natural frequency is $\hat{f}_{stru} = \sqrt{\hat{k}_{stru}/\hat{m}_{stru}}/2\pi$ (Hz). Note that this compensation model is only used in the analysis study, while the accurate model should be established or identified to compensate the real structural dynamics of the upper platform of the HIL simulator. Using the compensation model for the structural dynamics of the upper platform, the structural deformation $\hat{d}_{stru}(t)$ can be estimated from the estimated actual contact force $\hat{f}_{act}(t)$.

(2) Structural Dynamics Compensation of the Upper Platform

The structural deformation of the upper platform will lead to force error. The force error will further amplify the position error. To avoid the error accumulation and the simulation distortion, the contact force should be compensated. The structural dynamics based force compensation is proposed here.

The position error of the HIL simulator

$$\Delta p_{err}(t) = p_{des}(t) - p_{act}(t) \tag{7.10}$$

leads to the force error $\Delta f_{err}(t)$. The relationship between the position and force depends on the characteristics of the docking mechanisms

$$f_{act}(t) = z_{con}(p_{act}(t)) \tag{7.11}$$

In practice, the characteristics of the docking mechanisms do not change in one or several simulation steps. By real-time identifying the characteristics of the docking mechanisms, the compensation force can be obtained as follows

$$\Delta f_{stru}(t) = \left.\frac{\partial z_{con}}{\partial p_{act}}\right|_{p_{act}(t)} \hat{d}_{stru}(t) = -\hat{k}_{con}(t)\hat{d}_{stru}(t) \tag{7.12}$$

when $\hat{k}_{con}(t)$ is the estimated contact stiffness.

(3) Contact Stiffness Identification

In this study, the docking mechanisms are assumed to be stiffness dominant without the damping though the method can also be applicable to the damping case with minor revisions.

The actual relative position of the docking mechanisms can be estimated from

$$\hat{p}_{act}(t) = p_{sim}(t) - \hat{d}_{stru}(t) \tag{7.13}$$

where $p_{sim}(t)$ is the measured output position of the lower motion simulator and $\hat{d}_{stru}(t)$ is the estimated structural deformation of the upper platform.

From the Eq. 7.12, we have

$$\Delta f_{act}(t) = -k_{con}(t)\Delta p_{act}(t) \tag{7.14}$$

When a set of estimated actual position $\hat{p}_{act}(t)$,$\hat{p}_{act}(t-1),\cdots,\hat{p}_{act}(t-n)$ and the estimated actual contact force $\hat{f}_{act}(t)$,$\hat{f}_{act}(t-1),\cdots,\hat{f}_{act}(t-n)$ are given, the differences of the contact force or actual position at two time steps are

$$\Delta\hat{p}_{t-j} = \hat{p}_{act}(t-j) - \hat{p}_{act}(t-j-1), j = 1, ..., n \tag{7.15}$$

$$\Delta\hat{f}_{t-j} = \hat{f}_{act}(t-j) - \hat{f}_{act}(t-j-1), j = 1, ..., n \tag{7.16}$$

The time periods n is set to be a very small number so that the contact stiffness does not change. Then the contact stiffness is identified in real-time from $\{\Delta\hat{p}_{t-j}, \Delta\hat{f}_{t-j}\}_{j=1}^{n}$ using the recursive system identification methods.

7.4.4 Robot Dynamic Response Delay Compensation

A real-time response error based force compensation approach is used to compensate the dynamic response delay. The dynamic response model is not required. The response error of the lower motion simulator is

$$\Delta p_{resp}(t) = p_{des}(t) - p_{sim}(t) \tag{7.17}$$

This position error leads to a force error $\Delta f_{resp}(t)$. Similar to the structural dynamics compensation, by real-time identifying the contact stiffness of the docking mechanisms, the compensation force can be expressed by

$$\Delta f_{resp}(t) = \left.\frac{\partial z_{con}}{\partial p_{act}}\right|_{p_{act}(t)} \Delta p_{resp}(t) = -\hat{k}_{con}(t)\Delta p_{resp}(t) \tag{7.18}$$

where $\hat{k}_{con}(t)$ is the contact stiffness as in (7.12).

Finally, the compensated contact force can be obtained as follows

$$f_{comp}(t) = \hat{f}_{act}(t) + \Delta f_{stru}(t) + \Delta f_{resp}(t) \tag{7.19}$$

which is used to simulate the spacecraft motion at the next simulation step.

7.5 Verifications

In the simulations and experiments, the parameters such as the contact stiffness, contact natural frequency and initial velocity are similar to the real space contact. Specially, the contact stiffness is about tens or several hundred N/mm, the contact natural frequency is several Hz, and the initial velocity is about tens or one hundred mm/s.

7.5.1 Verifications by Software

To verify the performance of the compensation approach, the cases of the undamped and elastic contact are simulated. As an example, the contact stiffness is set as 100N/mm, the initial velocity is set as 100mm/s, the contact natural frequency is set as 5Hz. The time delay of the force measurement, the natural response frequency of the lower motion simulator and the structural natural frequency of the upper platform are selected similar to the real HIL system developed by us.

Firstly, assuming that there is only the force measurement delay, we study the performance of the phase lead compensation. Three cases of time delay: 1ms, 3ms and 5ms are studied as an example. Figure 7.7 shows the ideal velocity and the desired velocity with the phase lead compensation, where the CoR is near to one. Obviously, the phase lead compensation can prevent the HIL simulation divergence.

Secondly, assuming that there is only the dynamic response delay, we study the compensation performance of the real-time position error based force compensation. Three cases of the response frequency: 100Hz, 50Hz and 25Hz are simulated. Figure 7.8 compares the ideal and desired velocities with the delay compensation. The CoR after the compensation is very close to 1 for different cases of frequency response.

Thirdly, assuming that there is only the structural deformation of the upper platform, we study the compensation performance of the structural dynamics based force compensation. Figure 7.9 compares the ideal and desired velocities with the structural dynamics compensation at the 30Hz, 25Hz and 20Hz natural frequency and the 0.2 damping ratio of the structural dynamics. Figure 7.10 compares the ideal and desired velocities with the structural dynamics compensation at the 0.05, 0.1 and

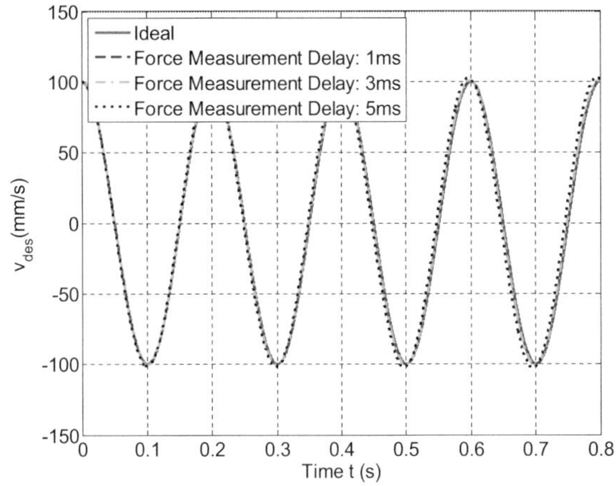

Fig. 7.7 Desired velocity with compensation of force measurement delay at different time delay

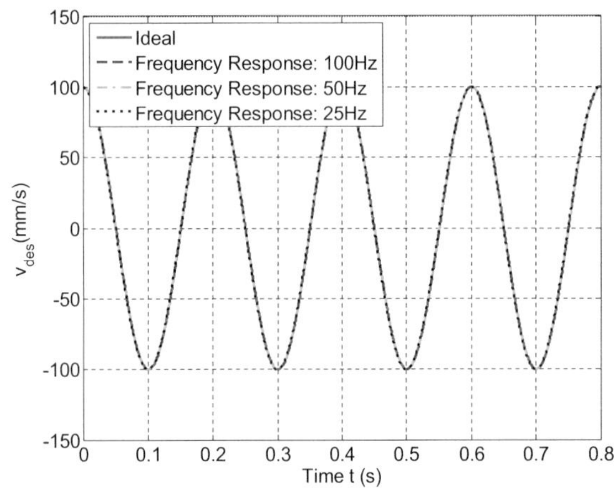

Fig. 7.8 Desired velocity with compensation of dynamic response delay at different response frequency

0.2 damping ratio and the 20Hz natural frequency of the structural dynamics. In this simulation, the CoR after the compensation is near to one.

Fourthly, we consider the force measurement system, lower motion simulator and upper platform together. In this case, the time delay is set as 5ms, the response frequency of the lower motion simulator is set as 25Hz, and the natural frequency of the structural dynamics of the upper platform is set as 20Hz. From Fig. 7.11, it can be concluded that without the compensation, the simulation divergence is very large, where the CoR is 1.9526. As seen in Fig. 7.12, only the force measurement delay compensation (FMDC) can reduce the simulation divergence, but still has a quite large divergence since the CoR after the compensation is 1.5950. From Fig. 7.13, it can be seen that by adding the dynamic response delay compensation (FMDC + DRDC), the system divergence is solved, but with a very small convergence (see the dash dotted green line). By further adding the structural dynamics compensation

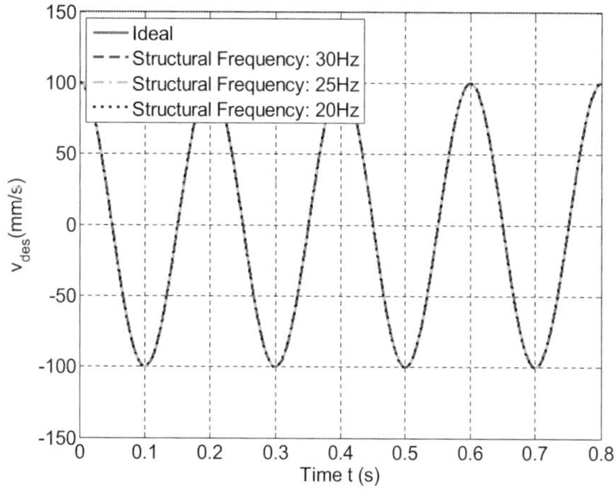

Fig. 7.9 Desired velocity with compensation of structural dynamics at different natural frequency

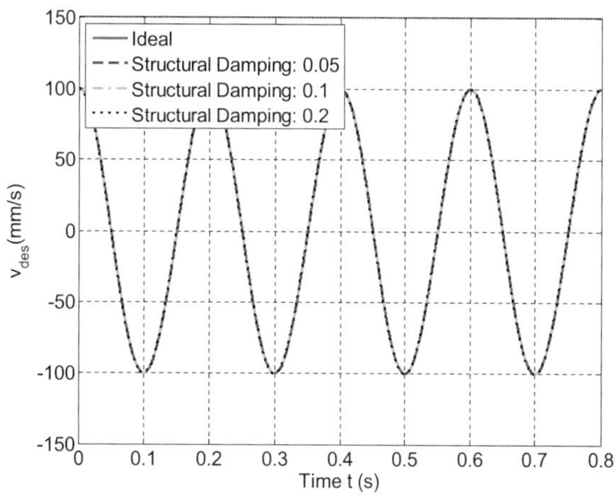

Fig. 7.10 Desired velocity with compensation of structural dynamics at different damping ratio

(FMDC + DRDC + SDC), the simulation convergence is handled and a satisfactory simulation performance is achieved, where the CoR is very close to 1 (see the dash black line). The proposed simulation distortion compensation approach is verified to be effective.

7.5.2 Verifications by Experiments

In the experiments, the developed HIL contact simulator is used. It is known that the small time delay of contact forces measuring, high frequency response of lower motion simulator, and high frequency of structural oscillations of upper platform

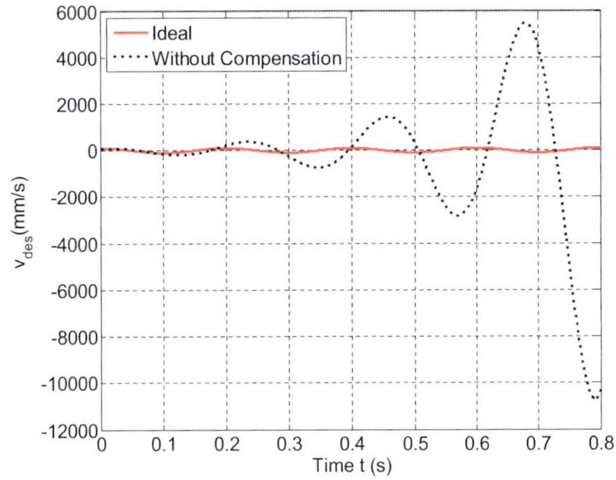

Fig. 7.11 Desired velocity without compensation

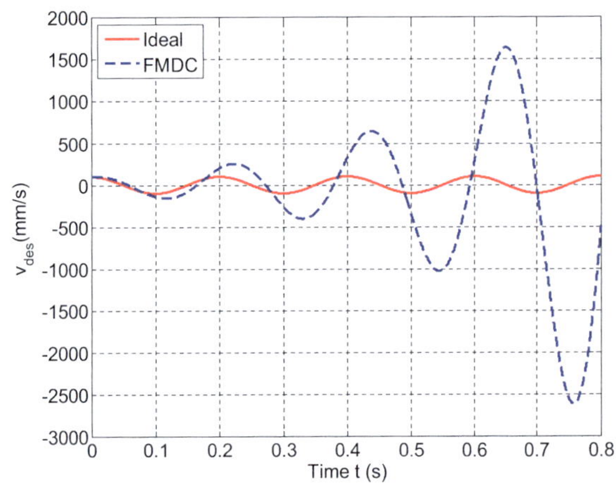

Fig. 7.12 Desired velocity with compensation of force measurement delay

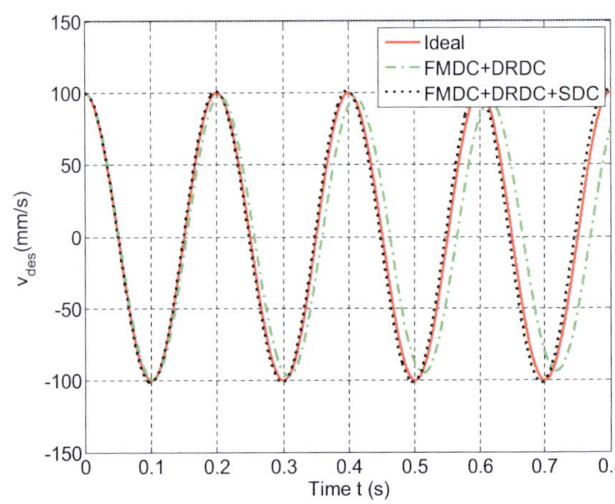

Fig. 7.13 Desired velocity with compensation of force measurement delay, dynamic response delay and structural dynamics

will make the HIL simulator system better. However, for our HIL system, the perfect design cannot be achieved due to current available hardware limitations.

Time interval between two measuring of contact force should be at least twice order less than the natural frequency of contact interaction of docking devices. In our study, considering the contact stiffness and spacecraft mass, the natural frequency of contact interaction of docking devices is less than 10Hz. Considering the data acquisition card, the contact force is measured at 20K Hz with 0.05ms time step. To reduce the force noise, the measured force is averaged for every 20 samples and then filtered for five averaged force samples. Thus the filtered force is got at 1K Hz. The simulation time step is set as 1ms, which is twice order less than the natural frequency of contact interaction of docking devices. Because five averaged force samples are filtered, about 2.5ms time delay exists. Thus, the force measuring time delay is not negligible in our case in order to get better simulation accuracy.

Contact stiffness of docking mechanism should be at least twice order less than lower motion simulator stiffness and upper platform stiffness. In our study, the contact stiffness of docking mechanism is about several hundred N/mm. The stiffness of the lower motion simulator and upper platform is about 10^4N/mm, which is twice order more than the contact stiffness of docking mechanism. If the contact stiffness k_{con} is growing, then the effective inertia and time constants of controlled drives of lower motion simulator could be increased. In our study, the contact stiffness does not change in a large range, thus the dynamic nature of the lower motion simulator only has a very small change, which can be ignorable.

In fact, the natural frequencies of the lower motion simulator and upper platform have significant influence on the HIL simulation accuracy. In theory, they should be twice order more than contact natural frequency. However, in practice their natural frequencies are very difficult to achieve twice order more than the contact natural frequency. Considering the inertia of the lower motion simulator and the actuation system capability, the natural frequency of the lower motion simulator is about 15Hz. Considering the inertia and stiffness of the upper platform, the natural frequency of the upper platform is about 30Hz. Therefore, the compensation for the position response error of the lower motion simulator and the structural deformation of the upper platform are necessary in order to get better simulation accuracy.

In our study, the oscillations d_{stru} of upper platform do not converge to zero before the beginning of next simulation step. Therefore, the structural deformation of upper platform cannot be ignorable in the compensation in order to get better simulation accuracy. At the beginning of the integration step of motion equations with computation of spacecraft kinematics and desired movement of lower motion simulator, the non-measured force values can be predicted using the phase lead method of the force measurement system according to the measured force values trend.

In the experiments, an undamped elastic rod is designed as a testing contact device because its ideal contact results can be easily calculated. The rod stiffness is calibrated and known before experiments. Four pictures in Fig. 7.14 show the main stages in the one contact: initial position, just contact, maximum position, and rebound. The

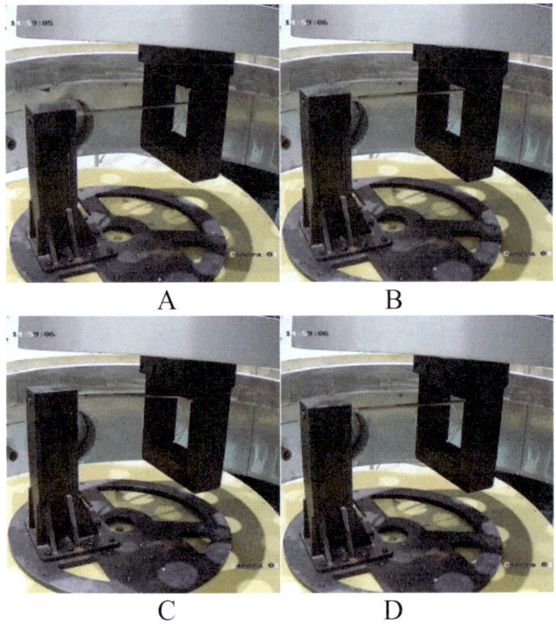

Fig. 7.14 Contact experiments

non-contact data between two frames are removed in order to show the figures more compactly.

Figure 7.15 shows the desired velocity at the 2.205Hz contact natural frequency, where the averaged CoR is about 1.1047, 1.0476, 0.9755 and 0.9936 for four cases respectively. Figure 7.16 shows the desired velocity without the compensation and with the compensation at the 3.298Hz contact natural frequency, where the averaged CoR is about 1.1887, 1.0830, 0.9800 and 0.9910 for four cases respectively. It can be concluded that without the compensation the simulation diverges significantly. With the force measurement delay compensation (FMDC), the simulation divergence is reduced, but it is still large. With the force measurement delay compensation and dynamic response delay compensation (FMDC + DRDC), the simulation divergence is reduced, but it becomes convergent. With the force measurement delay compensation, dynamic response delay compensation and structural dynamics compensation (FMDC + DRDC + SDC), the simulation becomes satisfactory with only a little convergence. In practice, a little convergence is preferred to guarantee the experimental safety as well as the accuracy.

The proposed method is applicable for the six-dimensional contact process. For the six-dimensional contact process, the distortion compensation problem is more challenging due to more complicated force measurement delay models, dynamic response errors and structural dynamics of upper platform. Experiments on real space docking mechanisms are also needed to verify the proposed method.

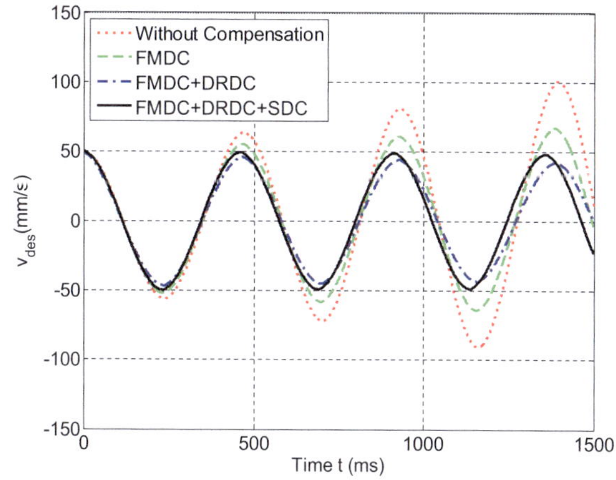

Fig. 7.15 Desired velocity with compensation in the experiment at 2.2Hz contact natural frequency

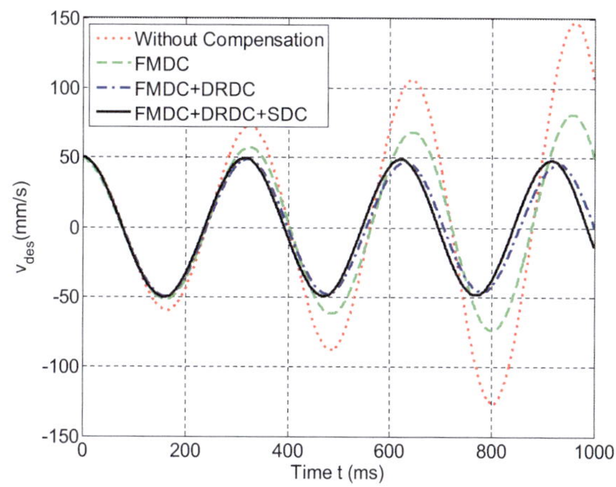

Fig. 7.16 Desired velocity with compensation in the experiment at 3.3Hz contact natural frequency

7.6 Summary

In this chapter, considering the force measurement delay, the dynamic response delay of the lower motion simulator and the structural dynamics of the upper platform, a HIL simulation distortion compensation approach is proposed. The force measurement delay is compensated by the phase lead, the dynamic response delay of the lower motion simulator is compensated by the position response error real-time feedback based force compensation, and the structural deformation of the upper platform is compensated by the structural dynamics based force compensation. Both simulation and experimental results show that the approach can compensate the simulation distortion and improve the simulation fidelity.

Reference

Qi, C. K., Gao, F., Zhao, X. C., Wang, Q., & Sun, Q. (2018). Distortion compensation for a robotic hardware-in-the-loop contact simulator. IEEE Transactions on Control Systems Technology, 26(4), 1170–1179.

Chapter 8
Distributed Distortion Compensation Control

8.1 Introduction

For a HIL simulator design, the main difficulty is the system instability and simulation divergence due to the subsystem time delay. Because the limitation of the actuation system bandwidth and mechanical system stiffness, the lower motion simulator generates the dynamic response delay of the actuation and structural deformation dynamics of the lower motion simulator during tracking the motion command. Similarly, the force measurement system including the force/torque sensor and noise filtering has the force measurement delay. The numerical simulation of dynamics model of space manipulator and target spacecraft has a numerical calculation delay. These delays in the system loop lead to HIL simulation error and system instability.

To overcome the time delay and then guarantee the stability and simulation accuracy, some delay compensators must be used in the simulation loop. Most existing delay compensation methods only consider one total time delay or only compensate one signal (i.e., force or motion), which are called lumped compensation here.

Because the force measurement delay, numerical calculation delay, dynamic response delay and structural deformation dynamics simultaneously exists in the system loop, all time delays should be considered in the compensation. If all blocks of HIL simulation system are linear dynamic system or pure time delay, then changing the order of blocks does not change the closed-loop characteristic equation. Therefore the delays can be integrated together into one delay in the modeling. In this case, methods that compensate one total time delay could be applicable. However, the robustness of these methods to the signal noise is reduced because the total time delay is larger. On the other hand, when subsystems between two time delays have nonlinear dynamics, the time delays cannot be summarized into one total time delay because the closed-loop characteristic equation is changed. In this case, the lumped time delay compensation cannot work well.

C. Qi et al., *Control of Hardware-In-The-Loop Space Contact Simulation Robot*,
https://doi.org/10.1007/978-981-95-1172-3_8

In this chapter, a distributed delay compensation approach is proposed for a HIL simulation system (Qi et al. 2022), where the force measurement delay, numerical calculation delay, dynamic response delay and structural deformation dynamics at different locations in the system loop are compensated in situ. Compared with lumped compensation, distributed compensation has stronger robustness to the signal noise and better adaptability to the varying contact natural frequency. Simulations and experiments under a rich set of conditions show that the proposed compensation approach can guarantee the system stability and simulation accuracy. The designed HIL simulator is successfully applied to the target capture experiment of space manipulator.

This chapter is arranged as follows. The HIL simulation system of space manipulator capture is introduced in Sect. 8.2. The distributed delay compensation approach is proposed in Sect. 8.3. Section 8.4 includes simulations and experiments. The last section gives a summary.

8.2 HIL Simulation System of Space Manipulator Capture

8.2.1 Space Manipulator and Capture Mechanism

The considered space manipulator is a 7-DOF manipulator, as shown in Fig. 8.1. The base of the space manipulator is the space station. The space manipulator uses an end-effector to capture the target adaptor on the space station or spacecraft. The end-effector is often a type of snare wires which is passively compliant to complete the soft capture. The target adaptor is often a bar with a shaft collar on its tip. The end-effector and the target adaptor are called active and passive capture mechanism respectively.

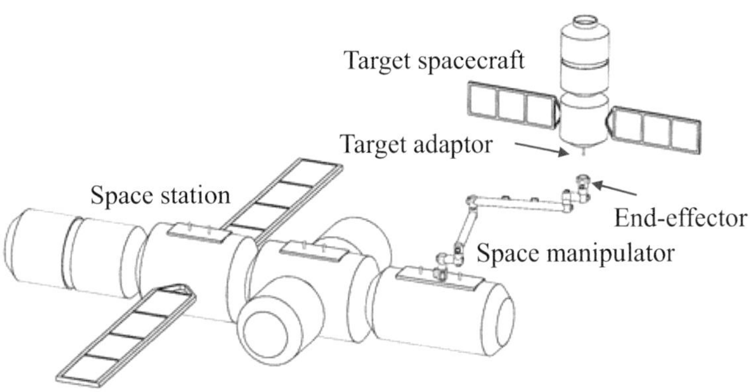

Fig. 8.1 Space manipulator capture system

8.2.2 Dynamics Models of Space Manipulator and Target

The dynamics models are used to compute the relative motion of the space manipulator and the target under the acted force/torque, which are desired motion of the lower motion simulator. The dynamics model of the space manipulator with a free-floating base (space station) can be described in the following form

$$\begin{aligned} &\begin{bmatrix} \mathbf{H}_b(\mathbf{x}_b, \boldsymbol{\theta}_m) & \mathbf{H}_{bm}(\mathbf{x}_b, \boldsymbol{\theta}_m) \\ \mathbf{H}_{bm}^T(\mathbf{x}_b, \boldsymbol{\theta}_m) & \mathbf{H}_m(\mathbf{x}_b, \boldsymbol{\theta}_m) \end{bmatrix} \begin{bmatrix} \ddot{\mathbf{x}}_b(t) \\ \ddot{\theta}_m(t) \end{bmatrix} \\ &+ \begin{bmatrix} \mathbf{C}_b(\mathbf{x}_b, \boldsymbol{\theta}_m, \dot{\mathbf{x}}_b, \dot{\theta}_m) & \mathbf{C}_{bm}(\mathbf{x}_b, \boldsymbol{\theta}_m, \dot{\mathbf{x}}_b, \dot{\theta}_m) \\ \mathbf{C}_{mb}(\mathbf{x}_b, \boldsymbol{\theta}_m, \dot{\mathbf{x}}_b, \dot{\theta}_m) & \mathbf{C}_m(\mathbf{x}_b, \boldsymbol{\theta}_m, \dot{\mathbf{x}}_b, \dot{\theta}_m) \end{bmatrix} \begin{bmatrix} \dot{\mathbf{x}}_b(t) \\ \dot{\theta}_m(t) \end{bmatrix} \\ &= \begin{bmatrix} \mathbf{F}_b(t) \\ \boldsymbol{\tau}_m(t) \end{bmatrix} + \begin{bmatrix} \mathbf{J}_b^T(\mathbf{x}_b, \boldsymbol{\theta}_m) \\ \mathbf{J}_m^T(\mathbf{x}_b, \boldsymbol{\theta}_m) \end{bmatrix} \mathbf{F}_e(t) \end{aligned} \tag{8.1}$$

where $\mathbf{H}_b \in \mathrm{R}^{6\times6}$ and $\mathbf{H}_m \in \mathrm{R}^{n\times n}$ are the inertia matrices of the base and the manipulator respectively, $\mathbf{H}_{bm} \in \mathrm{R}^{6\times n}$ is the coupling inertia matrix of the base and the manipulator, $\mathbf{x}_b \in \mathrm{R}^{6\times1}$ is the pose (position and attitude) of the base in the global frame, Euler angles are used for the attitude, $\boldsymbol{\theta}_m \in \mathrm{R}^{n\times1}$ represents the joint angles of the manipulator, $\mathbf{C}_b \in \mathrm{R}^{6\times6}$ and $\mathbf{C}_m \in \mathrm{R}^{n\times n}$ are the velocities dependent nonlinear terms of the base and the manipulator respectively, $\mathbf{C}_{bm} \in \mathrm{R}^{6\times n}$ and $\mathbf{C}_{mb} \in \mathrm{R}^{n\times6}$ are the velocities dependent nonlinear coupling terms of the base and the manipulator, $\mathbf{F}_b \in \mathrm{R}^{6\times1}$ is the force and torque acting on the center mass of the base, $\boldsymbol{\tau}_m \in \mathrm{R}^{n\times1}$ is the torque acting on the joints of the manipulator, $\mathbf{F}_e \in \mathrm{R}^{6\times1}$ is the contact force and torque acting on the manipulator end-effector, $\mathbf{J}_b \in \mathrm{R}^{6\times6}$ and $\mathbf{J}_m \in \mathrm{R}^{n\times n}$ are the Jacobian matrix dependent on the motion of the base and manipulator. Due to the large mass and inertia of the space station, the force acting on the end-effector of the space manipulator mainly changes the motion of the space manipulator and the target, but does not change the motion of the space station in a short time of target capture.

For capturing the free-floating target, the dynamics model of the target spacecraft can be described in the following form

$$\begin{aligned} \mathbf{M}_t\mathbf{a}_t &= \mathbf{F}_t \\ \mathbf{I}_t\dot{\omega}_t + \boldsymbol{\omega}_t \times \mathbf{I}_t\boldsymbol{\omega}_t &= \mathbf{T}_t \end{aligned} \tag{8.2}$$

where $\mathbf{M}_t \in \mathrm{R}^{3\times3}$ is the mass matrix of the target, $\mathbf{I}_t \in \mathrm{R}^{3\times3}$ is the inertia matrix of the target, $\mathbf{a}_t \in \mathrm{R}^3$ is the linear acceleration of the target, $\boldsymbol{\omega}_t \in \mathrm{R}^3$ is the angular velocity of the target, $\mathbf{F}_t \in \mathrm{R}^3$ and $\mathbf{T}_t \in \mathrm{R}^3$ are the force and torque acting on the center of mass of the target respectively. Note that for capturing the fixed target on the space station, the dynamics model of the target is not required.

The relative motion $\mathbf{P}_{cal}(t)$ of the space manipulator and target spacecraft is defined as the relative position and attitude of the end-effector frame of the space manipulator with respect to the target adaptor frame of the target spacecraft.

The position and attitude $\mathbf{x}_e$ of the end-effector frame of the space manipulator in the global frame is obtained from the pose $\mathbf{x}_b$ of the base in the global frame and the pose ${}^b\mathbf{x}_e$ of the end-effector of the space manipulator in the base frame. The pose ${}^b\mathbf{x}_e$ of the end-effector of the space manipulator in the base frame is obtained from the forward kinematics model ${}^b\mathbf{x}_e = \mathrm{FKM}(\boldsymbol{\theta}_m)$ of the space manipulator using joint angles $\boldsymbol{\theta}_m$.

The position and attitude $\mathbf{x}_t$ of the target adaptor frame of the target spacecraft in the global frame is obtained by integrating the Eq. 8.2 and considering the position and attitude of the target adaptor frame in the center of mass frame of the target.

8.2.3 Motion Simulator

The motion simulator is used to implement the relative motion of the space manipulator and the target. As shown in Fig. 8.2, in this study, the motion simulator consists of a 6-DOF parallel robot (lower motion simulator) and a 3-DOF parallel robot (upper platform). A 9-DOF setting has a large motion space that is used to cover both close approach and capture phases. The motion simulator is driven by closed-loop servo actuators. The position and attitude of the motion simulator are derived using the forward kinematics model based on the positions of actuation units from linear encoder. The kinematics model is calibrated in advance using the positions of actuation units from the linear encoder and the postures of the motion simulator from a 6-DOF laser tracker.

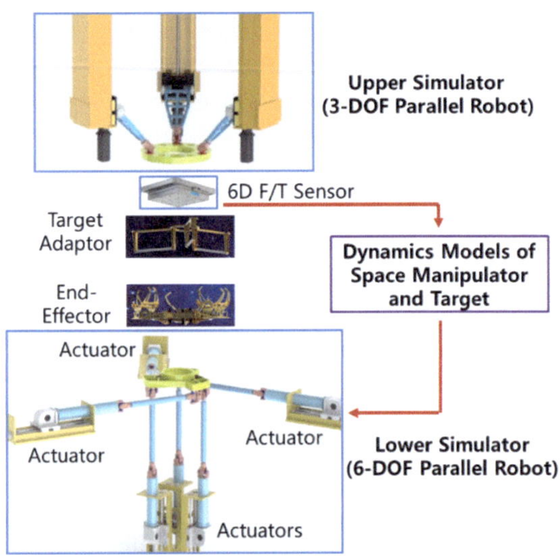

Fig. 8.2 Motion simulator for HIL simulation of space manipulator capture

8.2.4 Force/Torque Measurement

The contact force/torque of the capture process needs to be measured in real time as input of dynamics models of space manipulator and target. A 6-D force/torque sensor is installed behind the target adaptor to measure the force/torque acting on the target. Through a coordinate transformation, the force/torque acting on the end-effector of space manipulator can also be obtained. The sensor signal always has some noise, therefore a filter is used to reduce the noise level.

8.3 Distributed Delay Compensation Approach

8.3.1 Time Delay of Simulation Loops

Force Delay

The force delay comes from the force sensor, signal acquisition and noise filter. The measured contact force/torque $\mathbf{F}_{mea}(t) \in R^6$ is delayed to the actual contact force/torque $\mathbf{F}_{act}(t) \in R^6$. This delay is called force delay and can be approximated as a pure time delay

$$\mathbf{F}_{mea}(t) = \mathbf{F}_{\text{act}}(t - \tau_{\text{delay\_force}}) \tag{8.3}$$

where $\tau_{\text{delay\_force}}$ is the force delay time. This force delay can be estimated in advance using displacement-force experiments on rod-frame contact device.

Calculation Delay

Given the forces and torques on the right side of the Eq. 8.1, calculating the pose of the base and joint angles of the manipulator is complicated. In general, the multi-body dynamics model requires more calculation time than the single-body dynamics model. The calculation delay is the time taken to calculate dynamics models of space manipulator and target, which cannot be neglected. The calculated relative motion $\mathbf{P}_{cal}(t) \in R^6$ of the space manipulator and target spacecraft is delayed to the timely relative motion $\mathbf{P}_{tim}(t) \in R^6$ corresponding to the measured force $\mathbf{F}_{mea}(t)$

$$[\mathbf{P}_{tim}(t), \mathbf{V}_{tim}(t)] = \text{Dyn}(\mathbf{F}_{mea}(t)) \tag{8.4}$$

$$\mathbf{P}_{cal}(t) = \mathbf{P}_{tim}(t - \tau_{\text{delay\_}cal}) \tag{8.5}$$

where Dyn(•) denotes the dynamics model of space manipulator and target spacecraft. For simplicity, the calculation delay is approximated as a pure time delay and

τ_{delay_cal} is the calculation delay time which can be obtained in advance from the real-time computer clock. The ideal relative motion is obtained without the calculation delay.

Actuation Delay

The motion accuracy of the motion simulator depends on both actuators and parallel mechanism. The robot is driven by closed-loop servo actuators. The actuation delay is due to the dynamic response of the actuation system with a limited bandwidth. The commanded displacement of the actuators $\mathbf{D}_{cmd}(t)$ can be obtained using the inverse kinematics model IKM(·) of the motion simulator as follows

$$\mathbf{D}_{cmd}(t) = \text{IKM}(\mathbf{P}_{cal}(t)) \tag{8.6}$$

In general, the actuators have very similar dynamics, which can be modelled by a state-space model

$$\dot{\mathbf{X}}_i(t) = \mathbf{A}_i\mathbf{X}_i(t) + \mathbf{B}_i d_{cmd\_i}(t) \tag{8.7}$$

$$d_{act\_i}(t) = \mathbf{C}_i\mathbf{X}_i(t) \tag{8.8}$$

where $d_{cmd\_i}(t)$ is the commanded displacement of the ith actuator, $d_{act\_i}(t)$ is the actual displacement of the ith actuator, $\mathbf{X}_i(t) \in R^n$ is the state of the ith actuation closed-loop system. In general, n is designed to guarantee modelling accuracy. The above models of the actuator dynamics can be established in advance from the response experiments on the actuators. Amplitude error and phase delay of actuators exist because the bandwidth of the actuation system is always limited.

Mechanism Delay

The mechanism delay is due to the mechanism dynamics with a limited natural frequency. The parallel mechanism of the lower motion simulator including the links, joints and motion platform is outside the closed-loop servo actuators. The mechanism dynamics model is described as follows

$$\dot{\mathbf{Z}}(t) = \mathbf{G}\mathbf{Z}(t) + \mathbf{H}\begin{bmatrix} \mathbf{F}_{act}(t) \\ \mathbf{D}_{act}(t) \end{bmatrix} \tag{8.9}$$

$$\mathbf{P}_{act}(t) = \mathbf{W}\mathbf{Z}(t) \tag{8.10}$$

where $\mathbf{G}$, $\mathbf{H}$, $\mathbf{W}$ are the state matrix, input matrix and output matrix of the mechanism dynamics system, $\mathbf{Z}(t) \in R^m$ is the state of the mechanism dynamics system, $\mathbf{F}_{act}(t) \in R^6$ is the actual contact force/torque, $\mathbf{D}_{act}(t)=[d_{act\_1}(t), ..., d_{act\_6}(t)]^T$ is the actual displacement of the actuators, and $\mathbf{P}_{act}(t)$ is the actual motion of the motion simulator. In general, m is selected to guarantee the modelling accuracy. The above model of the mechanism dynamics can be identified in advance from the frequency response experiments on the parallel mechanism of the lower motion simulator. It is difficult to achieve very high natural frequency of parallel mechanism because the stiffness of each branch chain of parallel mechanism is limited. Therefore, the amplitude error and phase delay of mechanisms exist. Note that, here only the mechanism delay of the lower motion simulator is considered for simplicity. Using similar method, the mechanism dynamics of the upper platform can also be considered.

The constant assumption on the force delay and calculation delay is valid because these time delays are not changed in the developed HIL simulation system. In fact, the constant assumption on time delay is very common in the existing studies of HIL simulation systems. Of course, the time varying time delays can also be handled by several ways. The first way is to identify the time varying time delays in real-time and use the proposed compensation method. The second way is to establish a dynamic model describing the time varying time delays. The third way is to use a robust compensator design method when the bounds of time varying time delays are known. The use of dynamic models for actuation delay and mechanism delay are reasonable because these time delays are time varying in many HIL simulation systems including our developed one. The dynamic models are used in compensating the actuation delay and mechanism delay. When the dynamic models are inaccurate, the robust compensator design should be used to guarantee the simulation performance and system stability.

8.3.2 System Instability Due to Time Delay

The following simulation scenario is an undamped elastic contact between the end-effector and the target adaptor. The velocity curve of this case in space should be an oscillation with the same amplitude. Due to time delays of simulation system, the simulated velocity on the ground is an oscillation with some divergence. The singly delay simulation appear to be repetitive because multiple contacts are simulated for each delay in order to show the simulation divergence. One contact can be divided into the entry process and rebound process. The coefficient of the restitution (CoR), i.e., the ratio of the rebound velocity with the entry velocity, is used to describe the simulation divergence. As an example, the simulation parameters are set as: contact natural frequency 5 Hz, contact stiffness 50 N/mm, and initial velocity 100 mm/s. Other parameter settings do not change the influence trend of the time delay on the results.

The Effect of Force Delay

Firstly, we study the effect of the force delay on the HIL simulation by assuming that other system loops are ideal. Figure 8.3 shows HIL simulated relative velocity between space manipulator and target spacecraft with ideal (0 ms), 0.5 ms, 1 ms and 1.5 ms force delay. The corresponding CoR is 1.0, 1.025, 1.051 and 1.077 respectively. As shown in Fig. 8.4, when the force delay increases, the HIL simulation becomes more divergent. For higher contact natural frequency, the divergence is more significant.

The Effect of Calculation Delay

The effect of the calculation delay on the HIL simulation is similar to the effect of the force delay. When the calculation delay is larger, the HIL simulation becomes more divergent. For higher contact natural frequency, the divergence is larger. The results are ignored here.

The Effect of Actuation Delay

Then we study the effect of the actuation delay of the motion simulator on the HIL simulation. Figure 8.5 shows HIL simulated relative velocity with ideal, 10 Hz, 15 Hz and 20 Hz actuation bandwidth. The corresponding CoR is 1.0, 2.684,

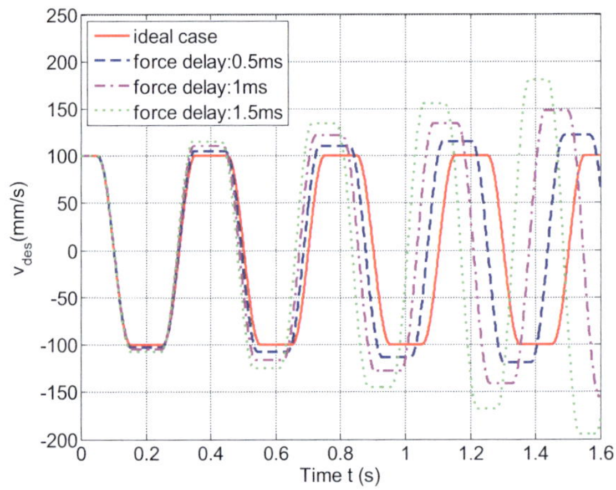

Fig. 8.3 HIL simulated relative velocity between space manipulator and target spacecraft with force delay

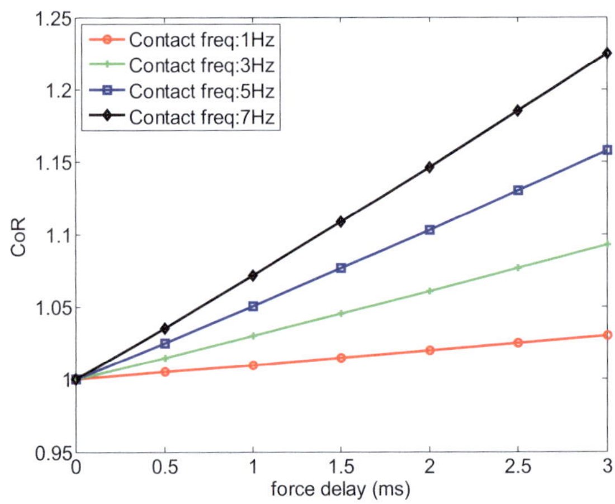

Fig. 8.4 CoR with force delay under different contact natural frequency

2.018 and 1.713 respectively. From Fig. 8.6, it can be seen that the HIL simulation becomes more divergent as the actuation bandwidth decreases. The simulation of the contact with a higher natural frequency is more challenging due to more divergence. In the software simulation, an actuation response action can be larger to simulate. However, for the implementation and experiment, the actuation velocity is limited by a maximum velocity provided by the actuator.

The Effect of Mechanism Delay

Now we study the effect of the mechanism delay on the HIL simulation. Figure 8.7 shows HIL simulated relative velocity with ideal, 15 Hz, 25 Hz and 35 Hz natural frequency. The corresponding CoR is 1.0, 1.038, 1.01 and 1.004. When the natural

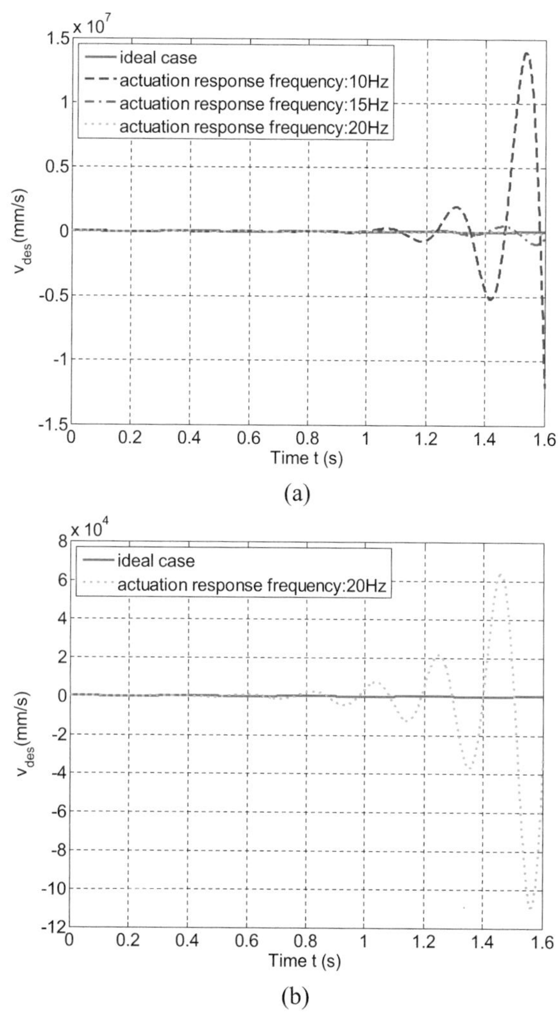

Fig. 8.5 HIL simulated relative velocity between space manipulator and target spacecraft with actuation delay

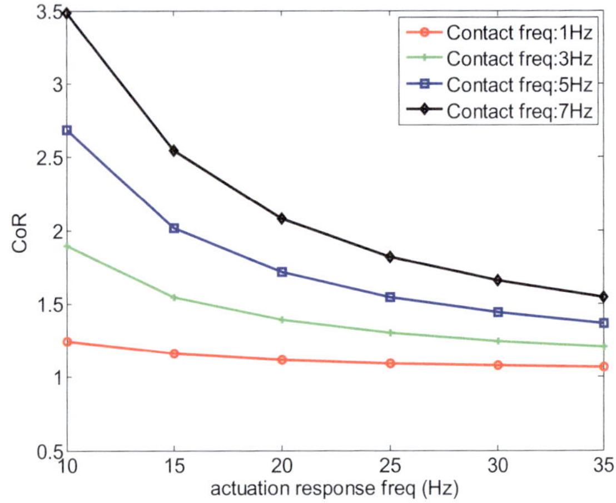

Fig. 8.6 CoR with actuation bandwidth under different contact natural frequency

frequency of the robot mechanism of the lower motion simulator is smaller, the HIL simulation becomes more divergent. The divergence trend with the delay is the same for different contact natural frequency. A higher contact natural frequency makes the divergence worse.

For the inner rules about each delay to system divergence, Fig. 8.8 is used to explain the divergence mechanism due to time delay from the perspective of phase analysis and work analysis. In the complex plane as shown in Fig. 8.8a, for the ideal elastic contact case (without time delay), the measured force F is orthogonal to the desired velocity V, and then the inner product of the vectors of measured force and desired velocity is zero, no external work is added to the system. The time delays in the simulation loop make the phase of the measured force F' delayed to the phase of the desired position P, and then the phase angle between the measured force F' and the desired velocity V is less than $\pi/2$ and external work is added to the system. One

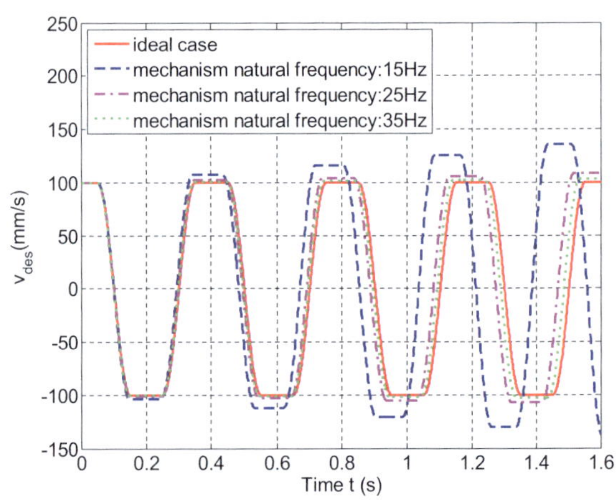

Fig. 8.7 HIL simulated relative velocity between space manipulator and target spacecraft with mechanism delay

contact can be divided into the entry process and rebound process. Due to the time delay, the work B done by the rebound force in the rebound process is larger than the work A done by the resistance force in the entry process as show in Fig. 8.8b. Moreover, the CoR becomes larger with more time delays.

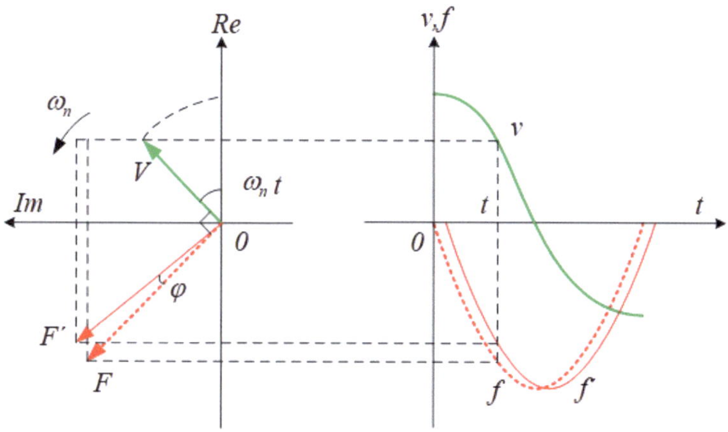

(a) Phase of measured force and desired velocity

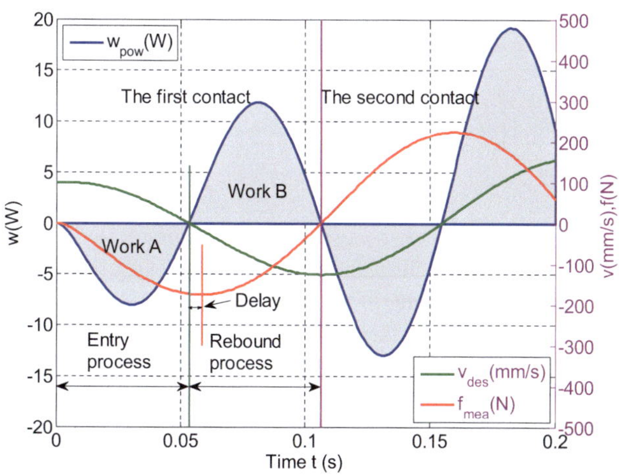

(b) Work analysis with time delay in the simulation loop

Fig. 8.8 Divergence mechanism due to time delay

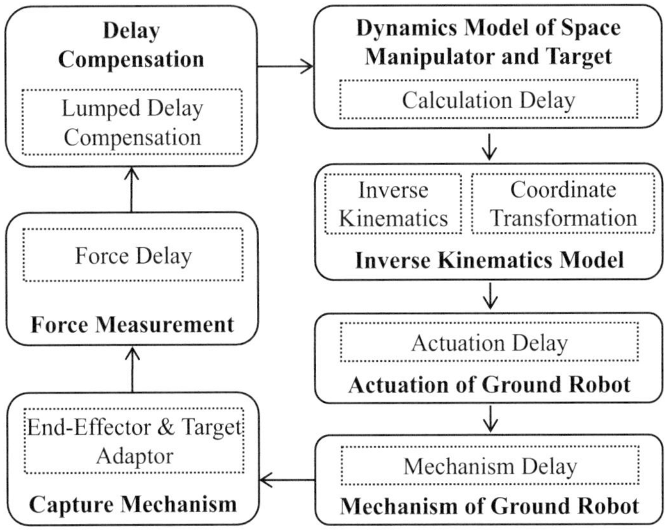

Fig. 8.9 Lumped delay compensation

8.3.3 Lumped Delay Compensation

If the force delay, calculation delay, actuation delay and mechanism delay can be integrated together, one lumped delay compensation can be used (Fig. 8.9). However, the lumped delay compensation requires a large phase lead or a high-order compensator, which is sensitive to the signal noise. When the subsystem has nonlinear dynamics, it is difficult to adjust the positions of all delays or dynamics to one position. Therefore, the application of lumped time delay compensation is limited.

8.3.4 Distributed Delay Compensation

In this study, a distributed delay compensation methodology (Fig. 8.10) is proposed for the HIL simulation system of space manipulator capture. Here the distributed delay compensation means that the delays are in different locations and the delay compensators are also in different locations. The force delay, calculation delay, actuation delay and mechanism delay are compensated in situ. The method does not require the assumption of linear systems. Distributed compensation could have stronger robustness to the signal noise because multiple small phase leads are required for distributed compensation while a large phase lead is required for lumped compensation.

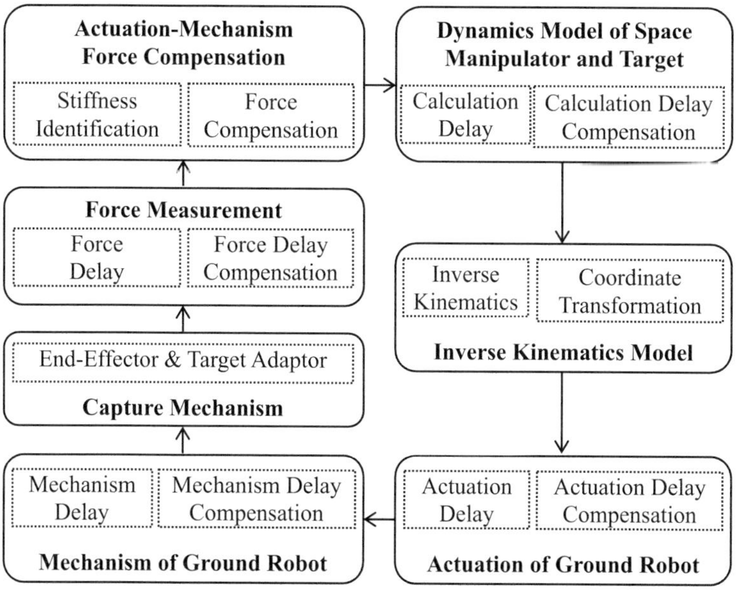

Fig. 8.10 Distributed delay compensation methodology

Force Delay Compensation

The force delay compensator is a phase lead compensator

$$\tilde{\mathbf{F}}_{mea}(t) = \mathbf{F}_{mea}(t) + \tau_{\text{lead\_force}}\mathbf{F}'_{mea}(t) \tag{8.11}$$

where $\mathbf{F}'_{mea}(t)$ is the derivative of the measured force. The compensated measured force $\tilde{\mathbf{F}}_{mea}(t)$ is expected to approach the actual contact force $\mathbf{F}_{act}(t)$. The derivative of the measured force is computed using the finite difference. Though the control period for the compensation is 1 ms, the sampling period for the force signal is 0.01 ms. After the force signal is averaged over 100 samples, the force noise is very small. Therefore, the derivative computation is robust to the measurement noise.

Calculation Delay Compensation

The calculation delay is compensated using the lead motion. The calculated relative motion $\tilde{\mathbf{P}}_{cal}(t)$ from $\mathbf{F}_{comp}(t)$ is compensated using the calculated relative velocity as follows

$$[\tilde{\mathbf{P}}_{cal}(t), \tilde{\mathbf{V}}_{cal}(t)] = \text{Dyn}(\mathbf{F}_{comp}(t)) \tag{8.12}$$

$$\overline{\mathbf{P}}_{cal}(t) = \tilde{\mathbf{P}}_{cal}(t) + \tau_{lead\_cal}\tilde{\mathbf{V}}_{cal}(t) \tag{8.13}$$

$\mathbf{F}_{comp}(t)$ is the compensated force after the force delay compensation and actuation-mechanism force compensation, which will approach the ideal force. The velocity $\tilde{\mathbf{V}}_{cal}(t)$ is obtained in the motion calculation and no additional calculation time is required. The compensated calculated relative motion $\overline{\mathbf{P}}_{cal}(t)$ is expected to approach the timely relative motion $\mathbf{P}_{tim}(t)$ corresponding to the compensated force $\mathbf{F}_{comp}(t)$.

Actuation Delay Compensation

The actuation delay is compensated using the lead command. The commanded displacement $\overline{\mathbf{D}}_{cmd}(t)$ of the actuators from the commanded motion $\overline{\mathbf{P}}_{cmd}(t)$ of the motion simulator is compensated using the displacement error $\Delta\mathbf{D}^{A}_{cmd}(t-\Delta t)$ in the last sampling time as follows

$$\overline{\mathbf{D}}_{cmd}(t) = \mathrm{IKM}(\overline{\mathbf{P}}_{cmd}(t)) \tag{8.14}$$

$$\Delta\mathbf{D}^{A}_{cmd}(t-\Delta t) = (\overline{\mathbf{D}}_{cmd}(t-\Delta t) - \mathbf{D}_{act}(t-\Delta t)) \tag{8.15}$$

$$\tilde{\mathbf{D}}_{cmd}(t) = \overline{\mathbf{D}}_{cmd}(t) + \Delta\mathbf{D}^{A}_{cmd}(t-\Delta t) \tag{8.16}$$

The actual displacement $\mathbf{D}_{act}(t)$ of actuation units is measured in real-time from linear encoder. The amplitude error and phase delay between the actual displacement $\mathbf{D}_{act}(t)$ and the commanded displacement $\overline{\mathbf{D}}_{cmd}(t)$ of the actuators can be reduced as much as possible.

Mechanism Delay Compensation

The mechanism delay is compensated using the mechanism dynamics model. The actual motion of the lower motion simulator can be estimated using the mechanism dynamics model

$$\dot{\hat{\mathbf{Z}}}(t) = \mathbf{G}\hat{\mathbf{Z}}(t) + \mathbf{H}\begin{bmatrix}\hat{\mathbf{F}}_{act}(t)\\ \mathbf{D}_{act}(t)\end{bmatrix} \tag{8.17}$$

$$\hat{\mathbf{P}}_{act}(t) = \mathbf{W}\hat{\mathbf{Z}}(t) \tag{8.18}$$

where $\hat{\mathbf{F}}_{act}(t)$ is the estimated actual contact force ($\hat{\mathbf{F}}_{act}(t) = \tilde{\mathbf{F}}_{mea}(t)$ here), $\hat{\mathbf{P}}_{act}(t)$ is the estimated actual motion of the lower motion simulator. Then the estimated motion error $\overline{\mathbf{P}}_{cal}(t-\Delta t) - \hat{\mathbf{P}}_{act}(t-\Delta t)$ in the last sampling time is compensated on the commanded motion of the lower motion simulator as follows

$$\Delta\mathbf{P}^{M}_{cmd}(t-\Delta t) = \overline{\mathbf{P}}_{cal}(t-\Delta t) - \hat{\mathbf{P}}_{act}(t-\Delta t) \tag{8.19}$$

$$\overline{\mathbf{P}}_{cmd}(t) = \overline{\mathbf{P}}_{cal}(t) + \Delta\mathbf{P}^{M}_{cmd}(t-\Delta t) \tag{8.20}$$

$\overline{\mathbf{P}}_{cmd}(t)$ is the commanded motion of the lower motion simulator. The amplitude error and phase delay between the actual motion $\mathbf{P}_{act}(t)$ and the compensated calculated relative motion $\overline{\mathbf{P}}_{cal}(t)$ are expected to be reduced.

Actuation-Mechanism Force Compensation

The actuation delay compensation can reduce the actuation displacement error $\overline{\mathbf{D}}_{cmd}(t) - \mathbf{D}_{act}(t)$, but this error cannot be zero because it is impossible to achieve an ideal actuator (i.e., the transfer function $\mathbf{G}(s) = \mathbf{I}$). And taking into account the mechanism error, finally there are always motion error between the actual relative motion $\mathbf{P}_{act}(t)$ of the simulator and the compensated calculated relative motion $\overline{\mathbf{P}}_{cal}(t)$. This part of motion error cannot be completely compensated using the position compensation. The relative motion error directly leads to the contact force error and then the simulation error. Here, an actuation-mechanism force compensation is used

$$\Delta\mathbf{P}^{M}_{cmd}(t) = \overline{\mathbf{P}}_{cal}(t) - \hat{\mathbf{P}}_{act}(t) \tag{8.21}$$

$$\Delta\mathbf{F}_{con}(t) = \mathbf{K}_{con}(t)\Delta\mathbf{P}^{M}_{cmd}(t) \tag{8.22}$$

$$\mathbf{F}_{comp}(t) = \tilde{\mathbf{F}}_{mea}(t) + \Delta\mathbf{F}_{con}(t) \tag{8.23}$$

The compensated force $\mathbf{F}_{comp}(t)$ approaches to the contact force corresponding to the compensated calculated motion $\overline{\mathbf{P}}_{cal}(t)$.

The distributed compensation algorithm is summarized in Fig. 8.11. By using the proposed distributed compensation, the HIL simulation accuracy is expected to be improved.

8.4 Verifications and Applications

8.4.1 Verifications by Software

(1) Force Delay Compensation Results

The following parameters are set as an example: contact natural frequency 5 Hz, contact stiffness 50 N/mm, and initial velocity 100 mm/s, and other parameter settings do not change the simulation conclusion. Three time delays: 0.5, 1 and 1.5 ms are studied. Figure 8.12 shows the HIL simulated relative velocity between space manipulator and target spacecraft. The force delay compensation can prevent the HIL simulation divergence when only a force delay exists.

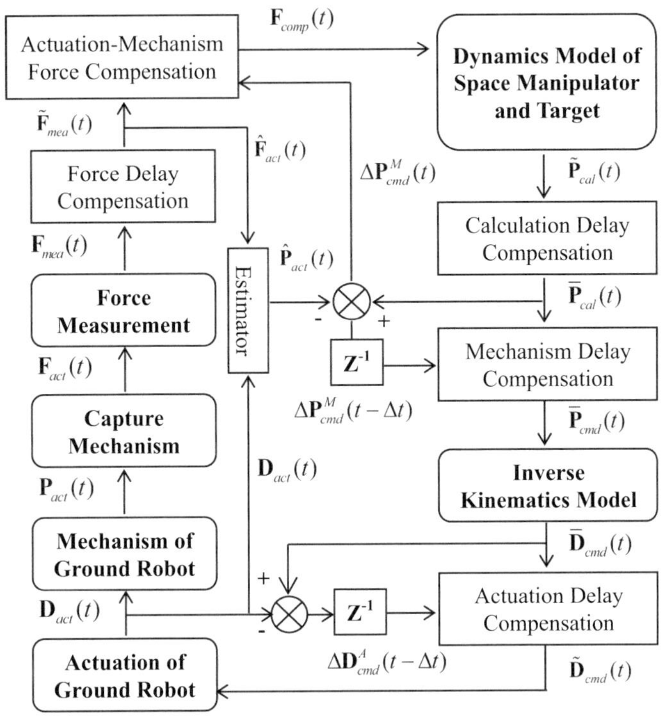

Fig. 8.11 Distributed delay compensation algorithm

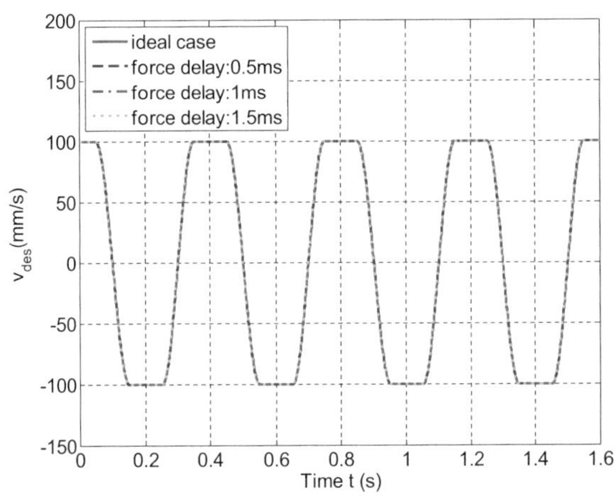

Fig. 8.12 HIL simulated velocity with FDC

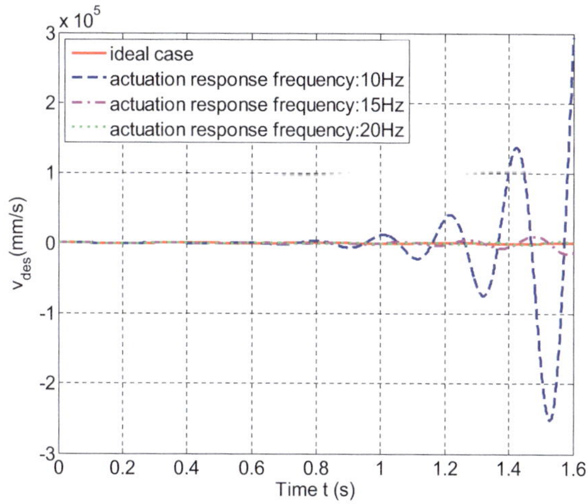

Fig. 8.13 HIL simulated velocity with ADC

(2) Calculation Delay Compensation Results

Similarly, when only a calculation delay exists, the calculation delay compensation works very well for different calculation values. The results are ignored here.

(3) Actuation Delay Compensation Results

When only an actuation delay exists, Fig. 8.13 shows the HIL simulated relative velocity with ideal, 10, 15 and 20 Hz actuation bandwidth. The corresponding CoR is 1.0, 1.827, 1.47 and 1.328 respectively. Compared with Fig. 8.5, the actuation delay compensation can largely reduce the HIL simulation divergence. However, it cannot completely solve the divergence problem because it is impossible to achieve an actuation system with the errorless tracking control.

(4) Mechanism Delay Compensation Results

When only a mechanism delay exists, Fig. 8.14 shows the HIL simulated relative velocity with ideal, 15 Hz, 25 Hz and 35 Hz natural frequency. The corresponding CoR is 1.0, 1.017, 1.005 and 1.002. Compared with Fig. 8.7, the mechanism delay compensation can reduce the simulation divergence. But it cannot completely prevent the simulation distortion.

(5) Distributed Delay Compensation Results

In this case, all delays are considered together. As shown in Fig. 8.15, without the compensation, the simulation divergence is large (CoR = 2.338). The force delay compensation (FDC) can reduce the divergence, but it is still quite large because the CoR is 2.249. By adding the calculation delay compensation (FDC + CDC), the simulation divergence is decreased, but with a large divergence (CoR = 2.076).

From Fig. 8.16, it can be seen that by adding the actuation delay compensation (FDC + CDC + ADC), the simulation divergence is reduced (CoR = 1.507). When

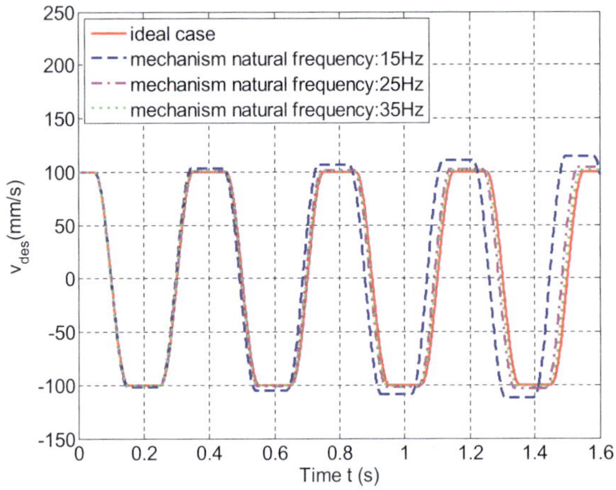

Fig. 8.14 HIL simulated velocity with MDC

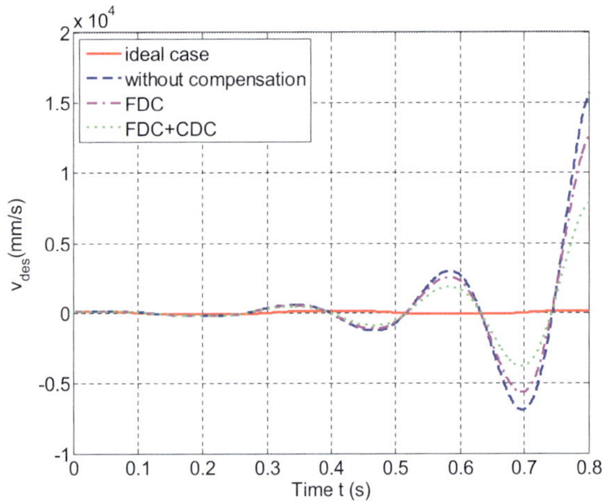

Fig. 8.15 HIL simulated velocity without compensation, with FDC, FDC + CDC

the mechanism delay compensation is further added (FDC + CDC + ADC + MDC), the simulation divergence still exists (CoR = 1.222).

As shown in Fig. 8.17, when the actuation and mechanism force compensation is further added (FDC + CDC + ADC + MDC + AMFC), an expected simulation performance is achieved, where the CoR (=1.0) is very close to 1. However, if the actuation delay compensation and mechanism delay compensation are removed (FDC + CDC + AMFC), the simulation divergence is prevented, but with a little convergence (CoR = 0.973) and a little wave for the rebound velocity. For the lumped force compensation, though it can prevent the velocity divergence, the unexpected and obvious wave of the rebound velocity exists. Figure 8.18 gives the coefficient of restitution (CoR) with different delay compensation approaches. The FDC + CDC + ADC + MDC + AMFC can achieve the best simulation accuracy (CoR = 0.9998).

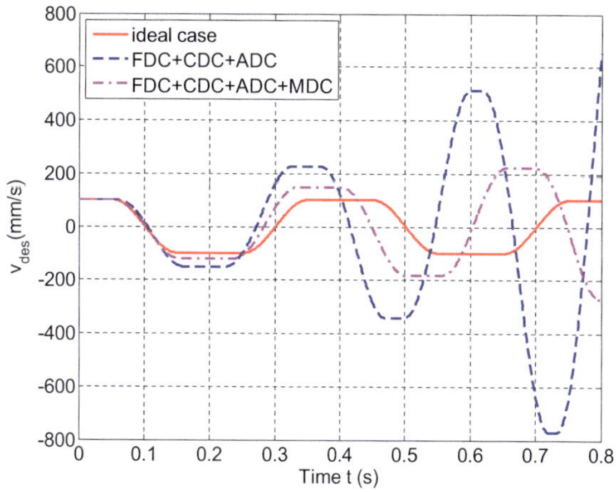

Fig. 8.16 HIL simulated velocity with FDC + CDC + ADC, FDC + CDC + ADC + MDC

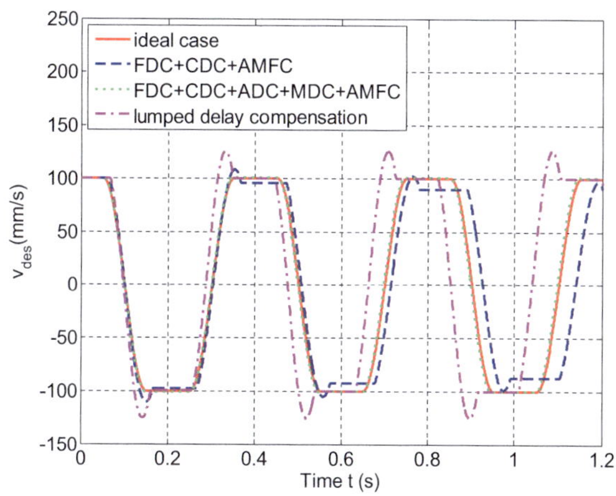

Fig. 8.17 HIL simulated velocity with FDC + CDC + AMFC, FDC + CDC + ADC + MDC + AMFC

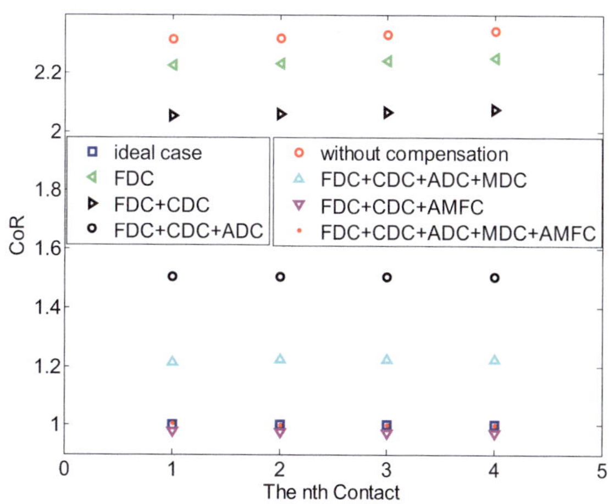

Fig. 8.18 CoR with delay compensation approaches

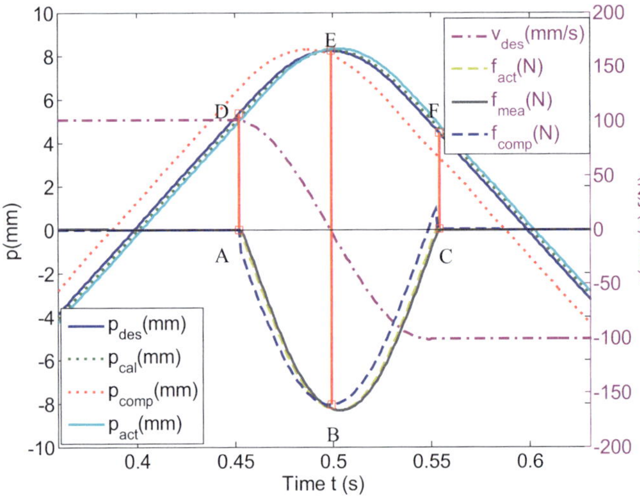

Fig. 8.19 Measured force, compensated force, calculated motion and compensated motion with FDC + CDC + ADC + MDC + AMFC

Phase Analysis and Work Analysis

Figure 8.19 provides the results with the FDC + CDC + ADC + MDC + AMFC for one contact process. Obviously, the measured force is delayed to the actual force, the calculated motion is delayed to the desired motion, and the actual motion is delayed to the compensated motion (i.e., commanded motion), while the compensated force is phase leaded to the measured force and the compensated motion is phased leaded to the calculated motion. The phase of the compensated force is close to the desired motion. That is why the simulation divergence is prevented with the FDC + CDC + ADC + MDC + AMFC.

In Fig. 8.19, the B and E correspond to the peak value of compensated force and the peak displacement of desired motion, A and C are starting and leaving time of one contact, D and F are desired positions at starting and leaving time of one contact. The work done by the resistance force from A to B is Win, where the motion is from D to E. The work done by the rebound force from B to C is Wout, where the motion is from E to F. Using the distributed compensation of FDC + CDC + ADC + MDC + AMFC, the ratio of Wout and Win is very close to 1. This is another explanation on why the simulation divergence is prevented with the FDC + CDC + ADC + MDC + AMFC.

Robustness to Force Signal Noise

Figures 8.20 and 8.19 compare the force signal noise effect on the lumped delay compensation and the proposed distributed delay compensation (FDC + CDC + ADC + MDC + AMFC). As shown in Fig. 8.20, for the lumped delay compensation the compensated force fcomp is very sensitive to the force signal noise due to its large

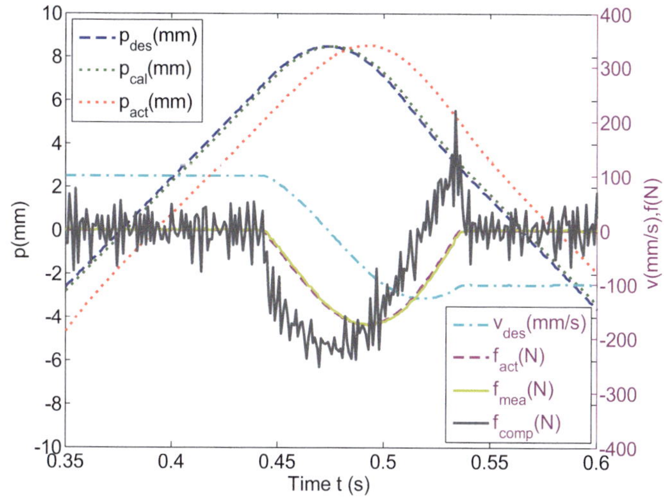

Fig. 8.20 Measured force, compensated force, calculated motion and desired velocity with lumped delay compensation

phase lead. As shown in Fig. 8.19, for the proposed distributed delay compensation, the compensated force is more robust to the force signal noise because multiple small phase leads are used.

Adaptability to Natural Frequency Varying Contact

Figure 8.21 compares the contact natural frequency effect on the lumped delay compensation and the proposed distributed delay compensation (FDC + CDC + ADC + MDC + AMFC). Note that the results in Fig. 8.17 are 5 Hz contact frequency, where both lumped delay compensation and distributed delay compensation are not divergent. For the 6 Hz contact in Fig. 8.21, it can be seen that the lumped delay compensation leads to significant divergence, while the proposed distributed delay compensation works well. This is because higher natural frequency contact leads to larger phase delay on the actuation and mechanism system, lumped delay compensation with fixed parameter is not adaptable to the natural frequency varying contact. However, the proposed distributed delay compensation is applicable for the natural frequency varying contact.

From the above simulations, it is verified that the proposed distributed delay compensation approach is very effective for the HIL simulation.

8.4.2 Verifications by Experiments

The experiments are conducted on the developed HIL simulator, where the developed motion simulator is shown in Fig. 8.22. To verify the proposed distributed delay

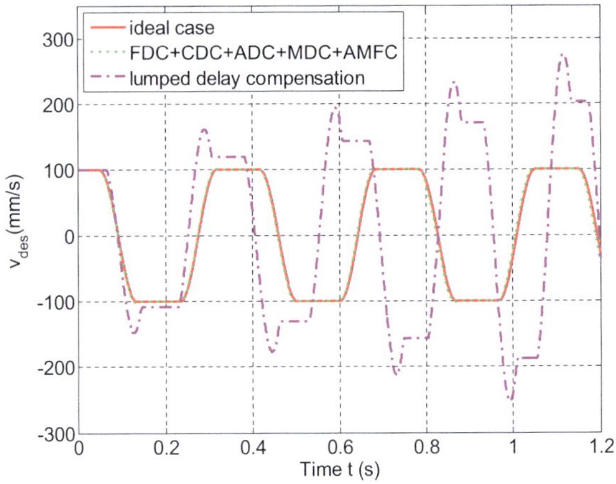

Fig. 8.21 HIL simulated velocity with FDC + CDC + ADC + MDC + AMFC, and lumped delay compensation at 6 Hz contact

compensation performance, a rod-frame contact experiment device in Fig. 8.23 is used, where it consists of an elastic rod and a highly rigid frame with a blade. Because the rod stiffness is known, the space capture results can be obtained analytically as the benchmark. There are two real-time computers which communicate through a real-time network. The numerical simulation computer performs the force acquisition, the numerical simulation of dynamics models, force delay compensation, actuation-mechanism force compensation and calculation delay compensation. The motion control computer performs the communication with the motor driver through the EtherCAT real-time bus, the motion control of the motion simulator, mechanism delay compensation and actuation delay compensation. The control period is 1 ms in the experiments. The block diagram of the simulation is given in Fig. 8.11.

Table 8.1 shows extensive experiments under a rich set of conditions: contact natural frequency from 1 to 8 Hz, contact stiffness from 48 N/mm to 750 N/mm, and initial velocity from 35 mm/s to 130 mm/s. The rebound velocity of the first five contacts are listed to obtain the simulation accuracy by CoR. As an example, Fig. 8.24

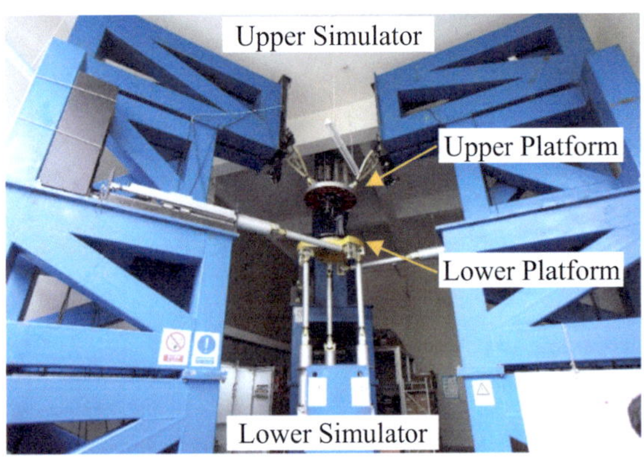

Fig. 8.22 The developed motion simulator for HIL simulator

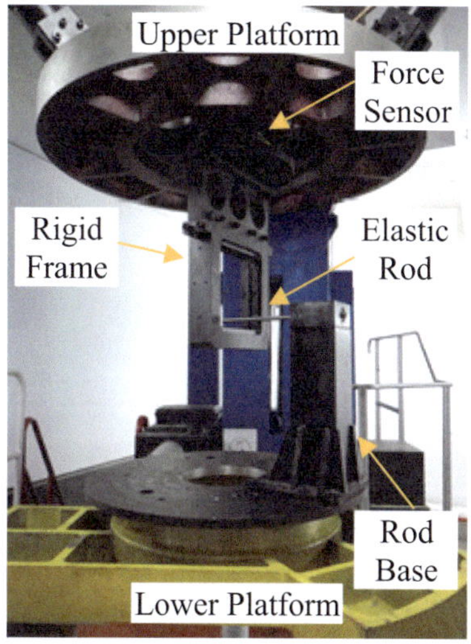

Fig. 8.23 Rod-frame contact experiment device

shows the HIL simulated velocity and force for the first three contacts in the rod-frame contact experiment under the contact natural frequency 3 Hz, contact stiffness 264.68 N/mm, and initial velocity 80 mm/s. The velocity direction is changed after each contact. The HIL simulation is a little convergent with a simulation accuracy of CoR = 96.10 ~ 99.18. From a large number of experiments, we can conclude that the proposed distributed delay compensation can guarantee the system stability and achieve the simulation accuracy better than 95%.

The force delay, calculation delay, actuation delay and mechanism delay all make the HIL simulation divergence. With larger delays, the simulation has more divergence. Both the force delay compensation and the calculation delay compensation work well for different delay values. The actuation delay compensation and the mechanism delay compensation cannot completely prevent the simulation distortion. The proposed distributed delay compensation includes the force delay compensation, calculation delay compensation, actuation delay compensation, mechanism delay compensation and actuation-mechanism force compensation, which can achieve good HIL simulation accuracy under a rich set of conditions. The phase analysis and work analysis explain why the proposed distributed delay compensation works.

8.4.3 Applications

The developed HIL simulator with the proposed distributed delay compensation has been successfully applied to simulate the target capture of space manipulator. Figure 8.25 shows the simulated relative velocity and contact force between the target and space manipulator during one space capture process. Figure 8.26 shows the simulated relative angular velocity and contact torque during the same space capture

Table 8.1 Experiments at different contact natural frequency, contact stiffness and initial velocity

Contact natural freq (Hz)	Contact stiffness (N/mm)	Initial vel (mm/s)	Rebound velocity after the ith contact (mm/s)					Simulation accuracy (CoR) (%)
			1st	2nd	3rd	4th	5th	
1	48.68	35	−33.75	32.34	−31.18	29.84	−28.65	95.72 ~ 96.42
2	75.75	−50	49.15	−48.69	48	−47.41	46.67	98.29 ~ 99.07
3	264.68	−80	79.11	−76.02	75.15	−74.42	73.80	96.10 ~ 99.18
4	301.50	−100	97.70	−96.74	94.96	−94.28	92.28	97.70 ~ 99.29
5	384.51	−130	125.31	−121.27	117.57	−115.54	112.2	96.39 ~ 98.28
6	470.59	−80	77.78	−76.94	74.69	−73.95	71.91	97.08 ~ 99.01
7	749.40	−60	57.90	−56.88	54.94	−54.34	52.97	96.50 ~ 98.90
8	688.40	−50	47.60	−46.36	44.91	−44.28	43.02	95.21 ~ 98.59

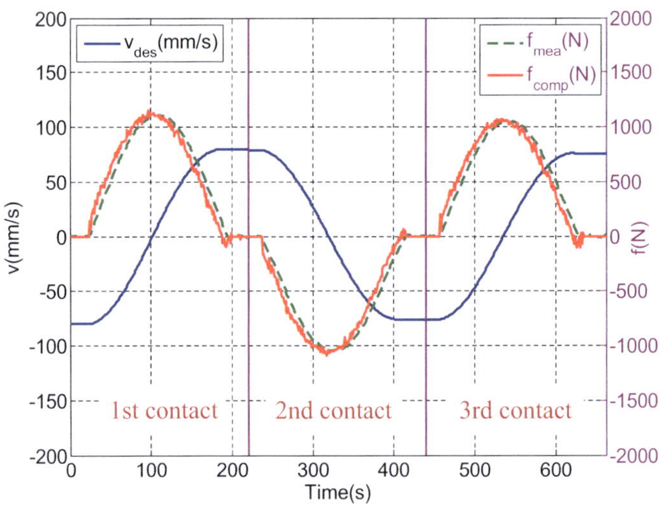

Fig. 8.24 HIL simulated velocity and force in the rod-frame contact experiment

process. It can be seen that the relative motion and force/torque gradually decrease, and quickly converge to the zero states after eight collisions. These experimental data can be used to verify the capture conditions, capture strategy, end-effector and target adaptor design and so on.

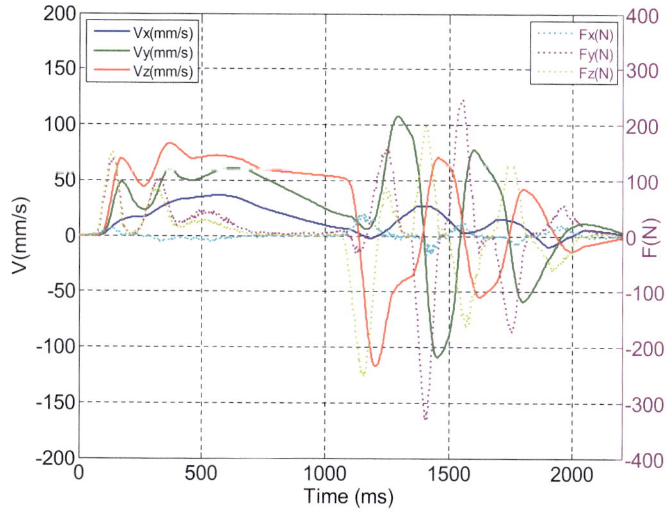

Fig. 8.25 HIL simulated velocity and force in target capture experiment of space manipulator

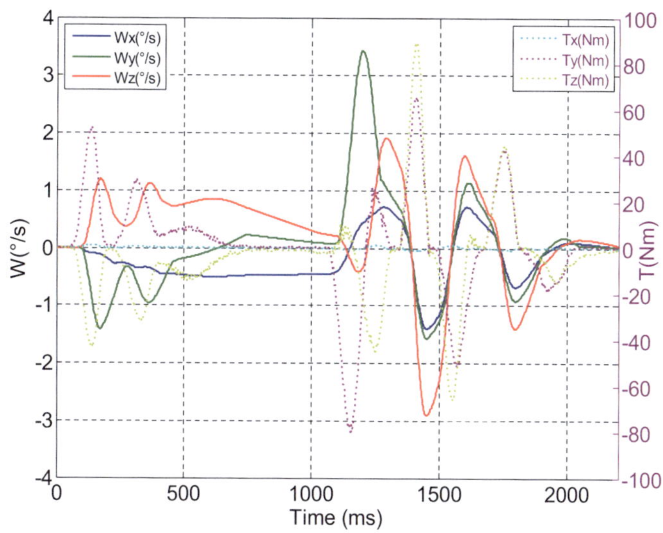

Fig. 8.26 HIL simulated angular velocity and torque in target capture experiment of space manipulator

8.5 Summary

In this chapter, a distributed delay compensation approach is proposed for the HIL simulation system. The force measurement delay, numerical calculation delay, dynamic response delay of the lower motion simulator and structural deformation dynamics of the lower motion simulator are compensated in situ. Compared with lumped compensation, distributed compensation has stronger robustness to the signal noise and better adaptability to the varying contact natural frequency. Simulations show that the proposed compensation approach can guarantee the system stability and simulation accuracy.

Reference

Qi, C. K., Li, D. J., Ma, W., Wei, Q. Q., Zhang, W. M., Wang, W. L., Hu, Y., & Gao, F. (2022). Distributed delay compensation for a hybrid simulation system of space manipulator capture. IEEE/ASME Transactions on Mechatronics, 27(4), 2367–2378.

Chapter 9
Learning Based Distortion Compensation Control

9.1 Introduction

Currently, although there are some compensation methods for phase delays and structural dynamics in the system loops of the HIL simulation system, they all require the system model or the model identification. In practice, the system model is not easy to be derived or identified. Therefore, in this case, a learning-based method will be a good way to design the delay compensator.

In this chapter, a learning-based distortion compensation method is proposed to compensate for phase delays and structural dynamics existing in the simulator (Qi et al. 2023). The measurement system delay of the contact force is compensated by a learning-based force compensator (LFC). For the motion simulator, the delay of the actuation system of the lower platform is compensated by a learning-based actuation compensator (LAC), and the structural dynamics of the lower and upper platforms are compensated by a learning-based structure compensator (LSC). The proposed learning-based method does not require the system models. Some simulations and experiments show that this method can reduce the HIL simulation error and achieve good accuracy.

In the following, Sect. 9.2 introduces the space docking HIL simulator. Section 9.3 presents the learning-based distortion compensation method. Section 9.4 includes the verifications. The final section gives a summary.

9.2 Space Docking HIL Simulator

Figure 9.1 shows the time delays and structural dynamics of HIL simulator. The upper platform and a docking mechanism (usually an active docking mechanism) form a local closed loop, while the lower platform and a docking mechanism (usually a passive docking mechanism) form a local closed loop. The whole system dynamics

C. Qi et al., *Control of Hardware-In-The-Loop Space Contact Simulation Robot*,
https://doi.org/10.1007/978-981-95-1172-3_9

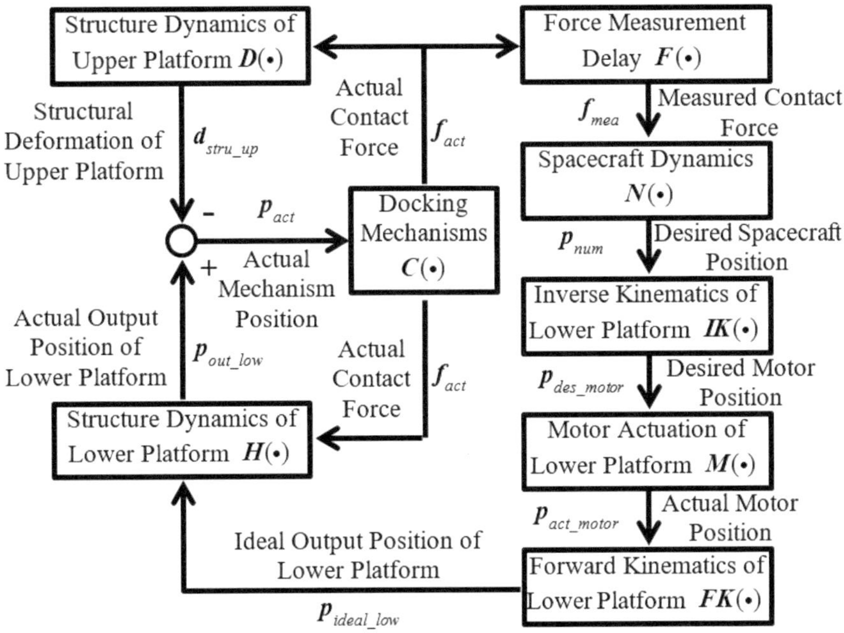

Fig. 9.1 Time delays and structural dynamics of HIL simulator

form a global closed loop (force measurement - > spacecraft dynamics - > actuation system of the lower platform - > lower platform of the motion simulator - > docking mechanisms with structural dynamics loops of upper platform and lower platform - > force measurement). The whole system is a multi-variable, multi-blocked and time delayed system. The whole system dynamics are very complex and difficult to be obtained. Therefore, in the following we mainly show the relationship between the system variables. Moreover, in order to simplify the equation derivations and expressions, we take one-dimensional case to show the method. The method can be applied to the six-dimensional case with more complicated equations.

9.2.1 *Contact Dynamic of Docking Mechanisms*

The deformation (i.e., impact depth) and contact characteristics (i.e., stiffness, damping and so on) of the docking mechanisms determine the actual contact force $f_{act}(t)$ as follows

$$f_{act}(t) = C(p_{act}(t)) \tag{9.1}$$

For an elastic damped contact, we have

$$f_{act}(t) = -k_{con}p_{act}(t) - b_{con}v_{act}(t) \tag{9.2}$$

where k_{con} and b_{con} are the stiffness and damping coefficients of docking mechanisms. $p_{act}(t)$ and $v_{act}(t)$ are the actual deformation and the actual deformation velocity of docking mechanisms respectively. The actual relative motion (i.e., actual deformation) $p_{act}(t)$ of docking mechanisms can be obtained as

$$p_{act}(t) = p_{out\_low}(t) - d_{stru\_up}(t) \tag{9.3}$$

where $p_{out\_low}(t)$ denotes the actual position outputted by the lower platform simulator, and $d_{stru\_up}(t)$ denotes the structural deformation caused by the upper platform.

9.2.2 Contact Force Measurement System Delay

The measurement system delay of the contact force is generated from the force sensor, signal processing and filtering. Considering all of them, the force measurement system delay is expressed by

$$f_{mea}(t) = F(f_{act}(t)) = f_{act}(t - \tau_{fms}) \tag{9.4}$$

where $f_{act}(t)$ and $f_{mea}(t)$ are the actual force and the measured force respectively, and τ_{fms} is the time delay. Note that the proposed distortion compensation control method is also applicable for other types of force measurement delay.

9.2.3 Numerical Simulation Model of Spacecraft Dynamics

The motion of the free-flying spacecraft changes with the docking contact force. $f_{num}(t)$ and $-f_{num}(t)$ are the force acted on the service spacecraft and the target spacecraft respectively. On the one hand, the dynamics model describing the motion $p_{ser}(t)$ of the service spacecraft under the contact force is represented by a differential equation

$$m_{ser}a_{ser} = f_{num}(t) \tag{9.5}$$

where m_{ser} is the mass of the service spacecraft, $a_{ser} = d^2p_{ser}/dt^2$ is the acceleration of the service spacecraft, $f_{num}(t)$ is the force acting on the service spacecraft.

On the other hand, the dynamics model describing the motion $p_{tar}(t)$ of the target spacecraft under the contact force is represented by

$$m_{tar}a_{tar} = -f_{num}(t) \tag{9.6}$$

where m_{tar} is the mass of the target spacecraft, $a_{tar} = d^2 p_{tar}/dt^2$ is the acceleration of the target spacecraft, $-f_{num}(t)$ is the force acting on the target spacecraft.

The relative position $p_{num}(t)$ between two spacecraft calculated from the numerical simulation models of spacecraft dynamics can be represented by

$$p_{num}(t) = p_{ser}(t) - p_{tar}(t) = N(f_{num}(t)) \tag{9.7}$$

The relative velocity $v_{num}(t)$ of two spacecraft can be obtained from $p_{num}(t)$. The natural frequency of the contact dynamic process during the spacecraft docking is an important parameter for the HIL simulation. It can be proved that for the direct impact in the docking process, the relative motion between two spacecraft equals the motion of an equivalent spacecraft with the mass $m_{spac} = m_{ser} m_{tar}/(m_{ser} + m_{tar})$. Then the natural frequency $freq_{con}$ (Hz) of the contact dynamic process is

$$freq_{con} = (1/2\pi)\sqrt{k_{con}/m_{spac}} \tag{9.8}$$

9.2.4 Actuation System Dynamics of Lower Platform

The 6-DOF parallel mechanism of the lower platform is driven by an actuation system of six servo-motors. The relative motion $p_{num}(t)$ between two spacecraft calculated from the numerical simulation models of spacecraft dynamics is the motion command of the lower platform. Using the inverse kinematic model $IKM(\cdot)$ of the lower platform, the desired motor position $p_{des\_motor}(t)$ of the actuation system can be obtained.

The response of the actuation system is expected to be fast enough in order to track the motion command dynamically. Due to a limited bandwidth, the actual motor position $p_{act\_motor}(t)$ of the actuation system is

$$p_{act\_motor}(t) = M(p_{des\_motor}(t)) \tag{9.9}$$

The natural undamped frequency $freq_{resp}$ (Hz) and the damping ratio ξ_{resp} are two important parameters to describe the dominant dynamics of the actuation system approximately.

9.2.5 Structural Dynamics of Lower Platform

For the actual output position $p_{act\_motor}(t)$ of the actuation system, using the forward kinematic model $FKM(\cdot)$ of the lower platform, the ideal output position $p_{ideal\_low}(t)$ of the lower platform can be obtained. However, the parallel mechanism of the

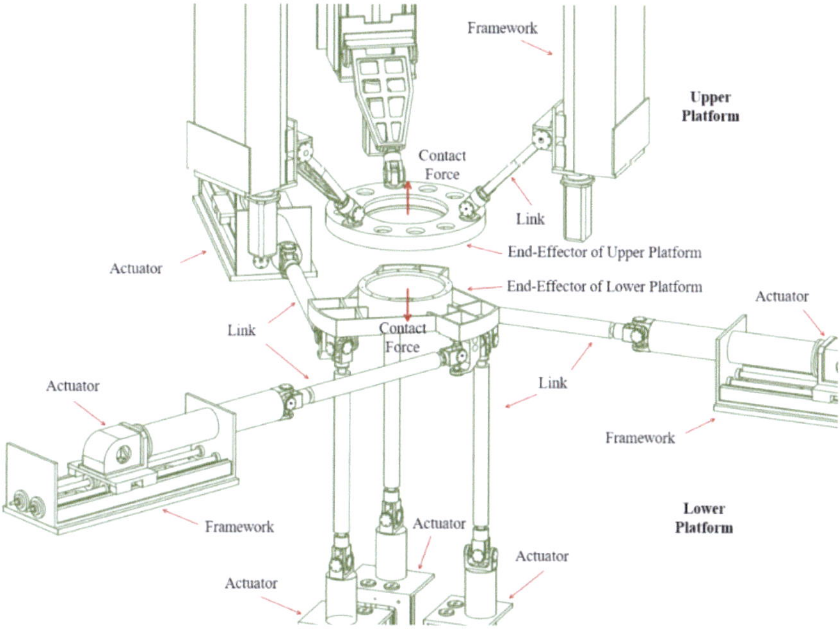

Fig. 9.2 Lower platform and upper platform

lower platform is not ideal due to the structural dynamics. The natural undamped frequency $freq_{stru\_low}$ (Hz) and the damping ratio ξ_{stru_low} are two important parameters to describe the structural dynamics of the lower platform approximately. The lower platform and upper platform are shown in Fig. 9.2.

Due to the structural dynamics of the parallel mechanism, the lower platform will have a structural deformation $d_{stru\_low}(t)$ under the contact force $f_{act}(t)$, that is

$$d_{stru\_low}(t) = D_{low}(f_{act}(t)) \tag{9.10}$$

This structural deformation will cause the relative position error between the upper platform and the lower platform.

Similarly, the dynamic output position p_{dyn_low} of the lower platform under the motor actuation will not equal the ideal output position p_{ideal_low} without the contact force

$$p_{dyn\_low}(t) = Q_{low}(p_{ideal\_low}(t)) \tag{9.11}$$

Therefore, the actual output position of the lower platform under the contact force and the motor actuation includes two terms

$$p_{out\_low}(t) = p_{dyn\_low}(t) + d_{stru\_low}(t) \tag{9.12}$$

9.2.6 Structural Dynamics of Upper Platform

The upper platform is a 3-DOF parallel robot, which is fixed in the space docking simulation in order to achieve a higher stiffness. The natural undamped frequency $freq_{stru\_up}$ (Hz) and the damping ratio ξ_{stru_up} are two important parameters of structural dynamics.

Although the platform actuation motor is locked, due to the structural dynamics of the mechanism, the platform end-effector will have a structural deformation $d_{stru\_up}(t)$ under the action of external contact forces $-f_{act}(t)$

$$d_{stru\_up}(t) = D_{up}(-f_{act}(t)) \tag{9.13}$$

Finally, the actual relative position $p_{act}(t)$ between docking mechanisms is

$$p_{act}(t) = p_{out\_low}(t) - d_{stru\_up}(t) \tag{9.14}$$

9.3 Learning-Based Distortion Compensation Approach

9.3.1 Learning-Based Distortion Compensation Framework

The proposed learning-based distortion compensation framework is shown in Fig. 9.3. It includes three learning-based distortion compensators. A Learning-based Force Compensator (LFC) is used to compensate for the measurement system delay of the contact force. A Learning-based Actuation Compensator (LAC) is used to compensate for the delay of the actuation system of the lower platform. A Learning-based Structure Compensator (LSC) is used to compensate for the structural dynamics of the lower and upper platforms. Through the three learning-based compensators, the phase delays and structural dynamics existing in the simulator are compensated. One compensator could be another way, but the network will be more complex. Using three compensators to compensate the unexpected dynamics locally can reduce the network complexity.

By simultaneously compensating for the commanded position and the measured contact force, the relative motion between the lower platform and the upper platform will approach to the ideal relative motion between two spacecraft, and the compensated contact force will be close to the ideal contact force of the space docking process.

The main advantage of this method is that the system models are not needed. Therefore, this method can be applied in the HIL simulation system where the system models are difficult to be obtained. In some cases, the phase delays and structural dynamics of the simulator could be partially known, then the method is also

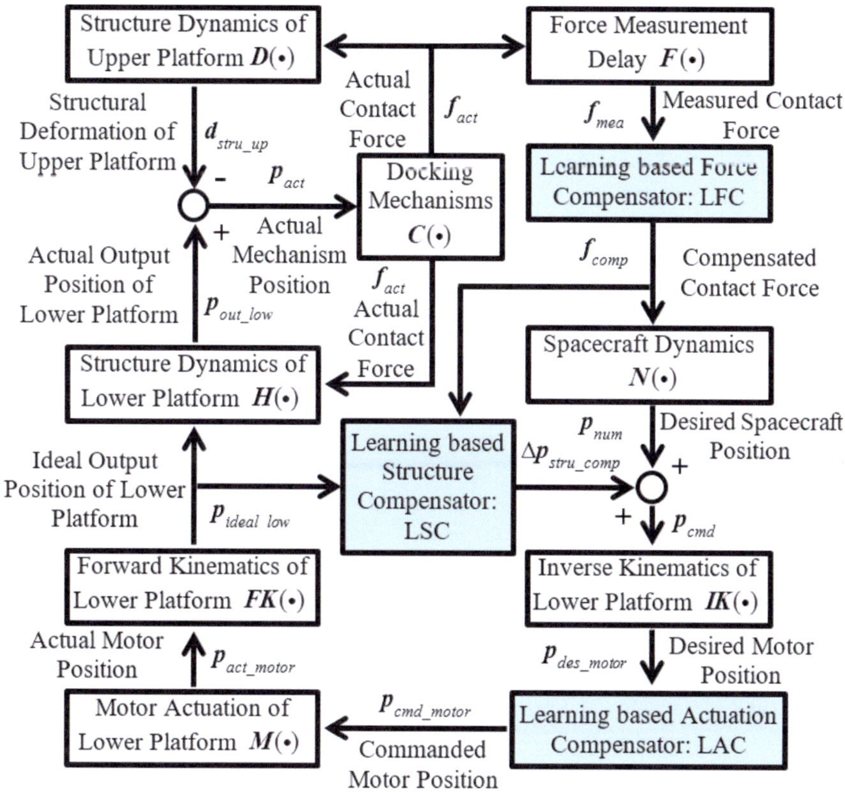

Fig. 9.3 Learning-based distortion compensation framework

applicable. The neural network compensators can become incremental compensators based on nominal compensators from partially known models.

9.3.2 *Learning-Based Force Compensator*

For the measurement system delay of the contact force, a force compensator is designed using a neural network. The input of the neural network is the measured contact force in the current and past few moments, and the output is the amount of the compensation force. The LFC is given as follows

$$\begin{aligned} f_{comp}(t) &= f_{mea}(t) + \Delta f_{mea}(t) \\ &= f_{mea}(t) \\ &\quad + NN_{LFC}(f_{mea}(t), f_{mea}(t-1), \ldots, f_{mea}(t-n)) \end{aligned} \tag{9.15}$$

where $\Delta f_{mea}(t)$ is the compensation force, $f_{comp}(t)$ is the contact force after the compensation (i.e., compensated force), and n is the history length of the measured force data. The parameter n cannot be too large or too small. It can be adjusted according to the compensation performance. From the perspective of the phase, the compensated contact force is ahead of the measured contact force. The network structure is a forward multi-layer neural network. The neurons are a set of Gaussian functions. The network weights are online trained using a contact process with known results.

9.3.3 Learning-Based Structure Compensator

For the structural deformation of the lower and upper platforms, a structure compensator is designed to add compensation on the desired spacecraft position

$$\Delta p_{stru\_comp}(t) = \Delta p_{low}(t) + \Delta p_{up}(t) \tag{9.16}$$

where $\Delta p_{low}(t)$ is the compensation displacement for the lower platform structural dynamics

$$\Delta p_{low}(t) = NN_{LSCL}(\Delta p_{low}(t-1), p_{ideal\_low}(t), f_{comp}(t)) \tag{9.17}$$

$\Delta p_{up}(t)$ is the compensation displacement for the upper platform structural dynamics

$$\Delta p_{up}(t) = NN_{LSCU}(\Delta p_{up}(t-1), f_{comp}(t)) \tag{9.18}$$

The neural networks are online trained using a contact process with known results.

The elastic deformations of the lower platform and the upper platform result in contact position errors. The learning-based structure compensator (LSC) adds the elastic deformation onto the numerical position

$$p_{cmd}(t) = p_{num}(t) + \Delta p_{stru\_comp}(t) \tag{9.19}$$

9.3.4 Learning-Based Actuation Compensator

For the delay of the actuation system of the lower platform, an actuation compensator is designed using a neural network. The input of the neural network is the desired motor position in the current and past few moments, and the output is the compensation amount of the commanded motor position

$$\begin{aligned} p_{cmd\_motor}(t) &= p_{des\_motor}(t) + \Delta p_{motor}(t) \\ &= p_{des\_motor}(t) \\ &\quad + NN_{LAC}(p_{des\_motor}(t), \ldots, p_{des\_motor}(t-q)) \end{aligned} \tag{9.20}$$

where $\Delta p_{motor}(t)$ is the compensation position, $p_{cmd\_motor}(t)$ is the commanded motor position after the compensation, and q is the history length of the desired motor position data. The parameter q cannot be too large or too small. It can be adjusted according to the compensation performance. The commanded motor position after the compensation is phase ahead of the desired motor position. This neural network is online trained using a motor position tracking process.

9.3.5 Learning Algorithms

The LAC can reduce the phase delay and the position error from the desired motor position $p_{des\_motor}(t)$ to the actual motor position $p_{act\_motor}(t)$. Because these two variables are known or can be measured directly, the LAC can be trained using a motor position tracking process. This individual training can reduce the training burden of three compensators.

The LFC can reduce the phase delay and the force error from the actual force $f_{act}(t)$ to the compensated force $f_{comp}(t)$. The compensated force is known, but the actual force is unknown, therefore the LFC cannot be trained individually.

The LSC can reduce the phase error and the position error from the desired spacecraft position $p_{num}(t)$ to the actual mechanism position $p_{act}(t)$. The desired spacecraft position is known, but the actual mechanism position cannot be measured directly, therefore the LSC cannot be trained individually.

After the LAC is trained using a motor position tracking process, the LFC and LSC are trained together using a contact process with known results. The optimization objective is to minimize the velocity error between the simulated velocity and the ideal velocity.

The feed-forward radial basis function neural network is used in the compensator design. The neurons are selected as Gaussian functions. The number of hidden layer nodes of the neural network is related to the complexity of the system dynamics to be compensated, which is adjusted by the compensation performance. In our simulations, $n = 5$ and $q = 5$ are enough to achieve good compensation performance.

Algorithm 1 Levenberg-Marquardt optimization method

Input: The initial network weights $\mathbf{W}$ and the optimization objective function $E(\mathbf{W})$

Output: The set of network weights $\mathbf{W}^*$ minimizing E

Steps:

$stop = 0;\ \nu = 2;\ \tau = 10^{-6};\ \varepsilon_1 = 10^{-12};\ \varepsilon_2 = 10^{-10}$

while (not stop)

    $\mathbf{A} = \mathbf{J}^T\mathbf{J};\ \mathbf{g} = \mathbf{J}^T E(\mathbf{W});\ \mu = \tau \times \max\{a_{ii}\}$

    Solve $(\mathbf{A} + \mu\mathbf{I})\boldsymbol{\delta} = -\mathbf{g}$

    if $\| \boldsymbol{\delta} \| \le \varepsilon_2 \times \| \mathbf{W} + \varepsilon_2 \|$

        $stop = 1$

    else

        $\mathbf{W}_{new} = \mathbf{W} + \boldsymbol{\delta};\ \rho = [E(\mathbf{W}) - E(\mathbf{W}_{new})] / [\boldsymbol{\delta}^T(\mu\boldsymbol{\delta} - \mathbf{g})]$

        if $\rho \le 0$

            reject $\mathbf{W}_{new};\ \mu = \mu \times \nu;\ \nu = 2 \times \nu$

        else

            accept $\mathbf{W}_{new};\ \mathbf{W} = \mathbf{W}_{new}$

            $\mu = \mu \times \max\{1/3, 1 - (2\rho - 1)^3\};\ \nu = 2$

            if $\| E(\mathbf{W}) - E(\mathbf{W}_{new}) \| < \varepsilon_1$

                $stop = 1$

Return $\mathbf{W}$

The training process of the compensator is actually a process of minimizing the optimization objective. During every single training step of the LFC and LSC, the set of all simulated velocity data $\mathbf{v}_{sim}$ and the set of all ideal velocity data $\mathbf{v}_{ideal}$ could be written as

$$\mathbf{v}_{\text{ideal}} = [v_{\text{ideal}_1}, v_{\text{ideal}_2}, \ldots, v_{\text{ideal}_{Ns}}]^T \tag{9.21}$$

$$\mathbf{v}_{\text{sim}} = [v_{\text{sim}_1}, v_{\text{sim}_2}, \ldots, v_{\text{sim}_{Ns}}]^T \tag{9.22}$$

where N_s is the number of sampling points during the simulation. The optimization objective E is set to be the variance between the simulated velocity and the ideal velocity

$$E = \frac{1}{2N_s} \sum_{k=1}^{N_s} (v_{\text{sim}_k} - v_{\text{ideal}_k})^2 \tag{9.23}$$

The Levenberg–Marquardt method (Hagan and Menhaj 1994) is used as an optimization algorithm for the training process. This method combines the advantages of the gradient descent method and the Gauss–Newton method. By calculating an evaluation criteria in every training step, it could adjust the iterative step in the optimization process so that it could avoid falling into a local extremum and accelerate the optimization process. The detailed algorithm is shown in Algorithm 1. Note that other kinds of neural networks and training methods can also be used here.

Some parameters of the algorithm are explained as follows. ε_1 and ε_2 are the pre-set threshold that stop the algorithm. v and τ are the pre-set parameters that determine how the iteration step length updates. $\mathbf{J}$ is the Jacobian matrix of the optimization objective, $\mathbf{g}$ is the gradient vector. a_{ii} is the eigenvalue of matrix $\mathbf{A}$. μ is the combination coefficient. $\boldsymbol{\delta}$ is the iteration step length of the optimization process. ρ could determine whether the iteration direction is correct or not.

9.4 Verifications

9.4.1 Verifications by Software

The advantage of software verifications is that the parameters of the simulation system, docking mechanisms, spacecraft and docking initial condition can be set easily. Typically, for the simulation system, the force measurement system delay is 3 ms, the actuation system natural frequency is 30 Hz, the lower platform mass is 200 kg, and the lower platform natural frequency is 30 Hz, the lower platform damping ratio is 0.05, the upper platform mass is 200 kg, and the upper platform natural frequency is 30 Hz, and the upper platform damping ratio is 0.05. For the docking mechanisms, the stiffness is 500 N/mm, the damping is set to be 0 N/(mm/s) for simplicity. The mass of the service spacecraft is 800 kg. The mass of the target spacecraft is 150 kg. The contact natural frequency is 10.01 Hz, which is derived from the contact stiffness and the equivalent mass of two spacecraft. For the docking initial condition, the initial relative velocity of two spacecraft is 80 mm/s. The above parameters are selected only as an example, and the use of other parameters do not change the conclusions.

The HIL simulation can give the relative velocity and the relative position between two spacecraft as well as the docking contact force. By comparing the relative velocity of two spacecraft before and after the contact, the divergence or convergence distortion level can be found. For one contact, the CoR is a key criterion to evaluate the distortion level. The CoR is defined as the ratio of the exit velocity v_{exit} to the entry velocity v_{entry}, i.e., $\mathrm{CoR} = v_{\mathrm{exit}}/v_{\mathrm{entry}}$. The exit velocity and the entry velocity can be obtained from the velocity curve after and before the contact. For an undamped elastic contact process, due to the energy conservation the relative velocity of two spacecraft should be an ideal cosine curve, that is $\mathrm{CoR} = 1$.

As shown in Fig. 9.4, the HIL simulation has a very large divergence (CoR = 3.5941) if no distortion compensator is applied. Figure 9.5 shows several velocity curves at different number of training cycles. It can be seen that the distortion compensator works better with more training cycles and the training process is convergent. As shown in Fig. 9.6, the learning-based force compensator (LAC) can reduce the divergence, but the divergence is still unacceptable (CoR = 1.3370). By adding the learning-based actuation compensator (LFC + LAC), the system divergence is further reduced (CoR = 1.1617). By further adding the learning-based structure compensator (LFC + LAC + LSC), the most desired slightly convergent result is achieved (CoR = 0.9937). Figure 9.7 shows the CoR of the HIL simulation with the above compensation methods.

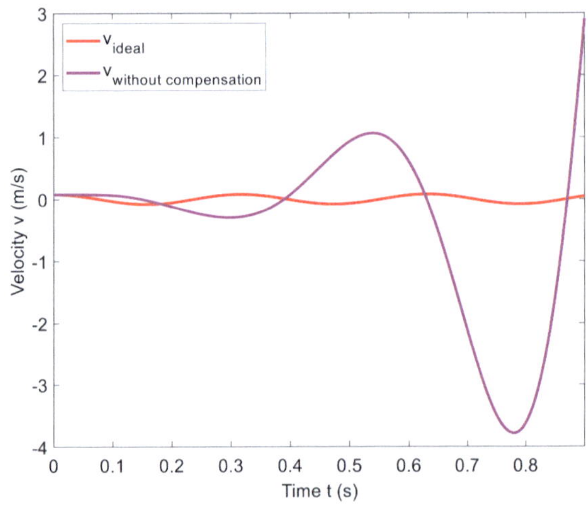

Fig. 9.4 Velocity without compensation

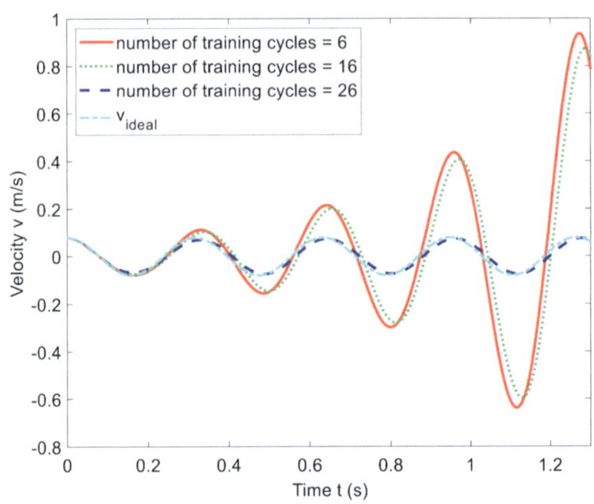

Fig. 9.5 Velocity with LFC + LAC + LSC in the training process

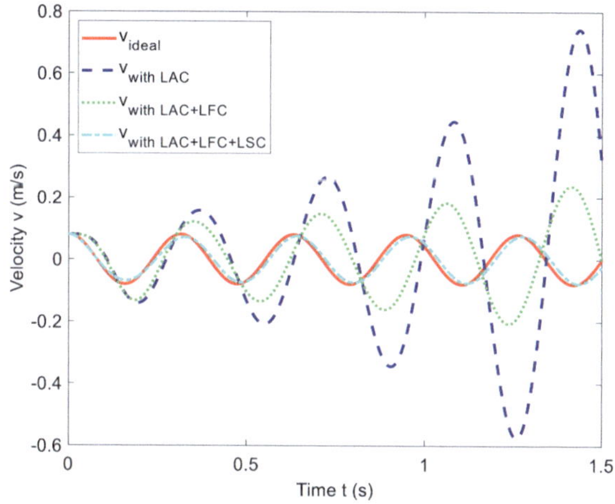

Fig. 9.6 Velocity with different compensation methods

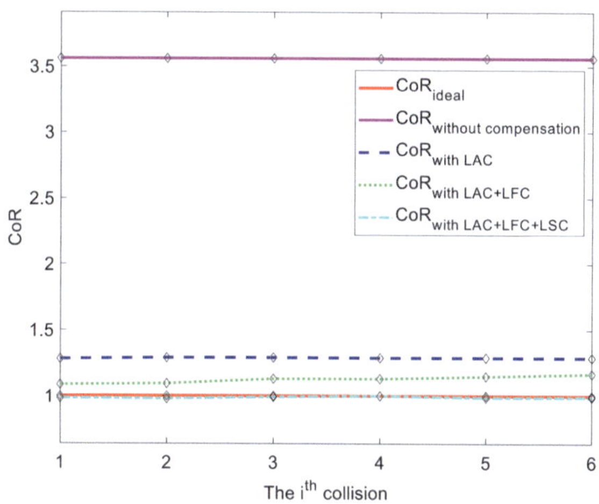

Fig. 9.7 CoR with different compensation methods

Figure 9.8 gives the desired motor position, the commanded motor position and the actual motor position simulated with the LAC. The actual motor position lags behind the commanded motor position, while the actual motor position is matched with the desired motor position. This means that the objective of the LAC is achieved. Figure 9.9 illustrates the actual, measured and compensated forces simulated with the LFC + LAC + LSC. The measured force lags behind the actual force, while the compensated force almost matches the actual force. This means that the objective of the LFC is achieved. Figure 9.10 shows the desired and commanded spacecraft positions, and the actual mechanism position simulated with the LFC + LAC + LSC. There are some position errors between the actual mechanism position and the commanded spacecraft position, while the actual mechanism position is consistent with the desired spacecraft position. This means that the objective of the LSC is achieved. The above analyses explain why the proposed method works.

The HIL simulations are also conducted under different system parameters and simulation parameters. In Table 9.1, the difference between the ideal CoR and the actual CoR represents the simulation error. The obtained results show that the proposed method (LFC + LAC + LSC) works very well under different parameters.

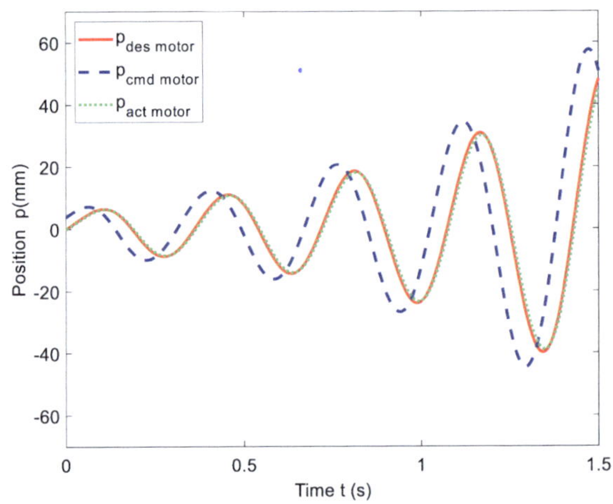

Fig. 9.8 Desired, commanded and actual motor position with LAC

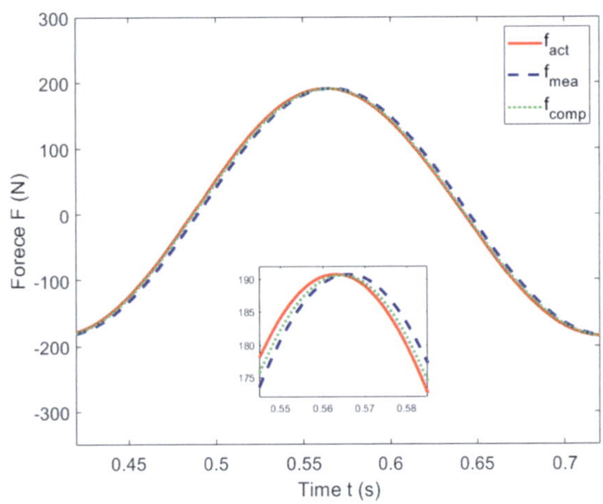

Fig. 9.9 Actual, measured and compensated force with LFC + LAC + LSC

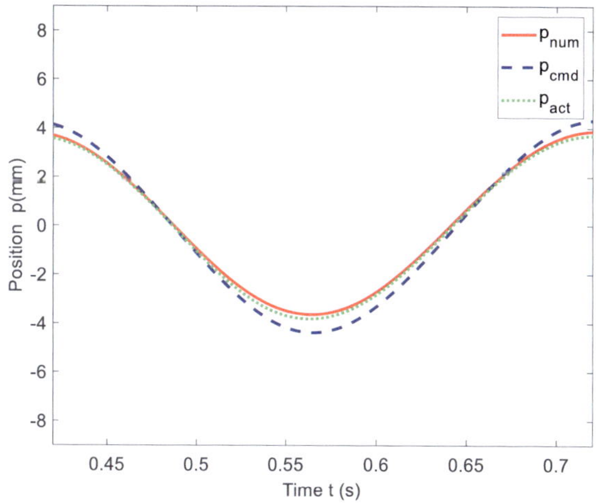

Fig. 9.10 Desired and commanded spacecraft position, and actual mechanism position with LFC + LAC + LSC

9.4.2 *Verifications by Experiments*

Experiment verifications are conducted on our developed space docking HIL simulator in Fig. 9.11. The contact experiment is performed on a test device as shown in Fig. 9.12. The elastic rod has the known stiffness and zero damping. The stiffness of lower and upper blades of the rigid frame is very large so that the contact between the rod and the lower (or upper) blade only causes the deformation of the elastic rod. During the contact process, the rod moves up and down in the rigid frame. The contact result between the rod and the blade in space is known, and for example, the velocity curve is an ideal cosine curve. Therefore, after three distortion compensators are trained, the contact experiment on this device can be used to verify the LFC + LAC + LSC method.

Figure 9.13 shows the simulated velocity for a 1 Hz contact with an initial velocity 20 mm/s, we can see that without the distortion compensation, the HIL simulation is divergent significantly (CoR = 1.055); and the LFC + LAC + LSC method can achieve high fidelity of convergent simulation (CoR = 0.973). Figure 9.14 shows the simulated velocity for a 4 Hz contact with an initial velocity 60 mm/s. In this case, the HIL simulation without the distortion compensation is more divergent (CoR = 1.160); and the LFC + LAC + LSC method can make the HIL simulation divergent (CoR = 0.934). In practice, several percentage points of convergence error is satisfactory for space docking experiments.

Table 9.1 HIL simulations by software under different system and simulation parameters

Parameters		Cases			
		A	B	C	D
Simulation parameters	Mass of service spacecraft (kg)	800	800	800	800
	Mass of target spacecraft (kg)	150	150	150	150
	Contact stiffness (N/mm)	200	300	400	500
	Contact damping (N/(mm/s))	0	0.01	0.02	0.04
	Contact natural frequency (Hz)	6.33	7.76	8.96	10.01
	Docking initial velocity (mm/s)	40	60	80	100
System parameters	Force measurement system delay (ms)	5	4	3	2
	Actuation system natural frequency (Hz)	10	10	20	20
	Lower platform mass (kg)	200	200	200	200
	Lower platform natural frequency (Hz)	20	20	30	30
	Lower platform damping ratio	0.05	0.05	0.1	0.1
	Upper platform mass (kg)	200	200	200	200
	Upper platform natural frequency (Hz)	20	20	30	30
	Upper platform damping ratio	0.05	0.05	0.1	0.1
CoR	Ideal	1	0.999	0.999	0.997
	Without compensation	5.896	5.747	4.316	4.499
	LAC	1.575	1.497	1.413	1.427
	LFC + LAC	1.253	1.204	1.179	1.172
	LFC + LAC + LSC	0.997	0.978	0.986	0.984

9.5 Summary

In this chapter, the distortion compensation problem for a space docking HIL simulator with unknown system loop models is studied. A learning-based distortion compensation method is proposed. The measurement system delay of the contact force is compensated by a learning-based force compensator (LFC). For the motion simulator, the delay of the actuation system of the lower platform is compensated by a learning-based actuation compensator (LAC), and the structural dynamics of the

Fig. 9.11 Space docking HIL simulator

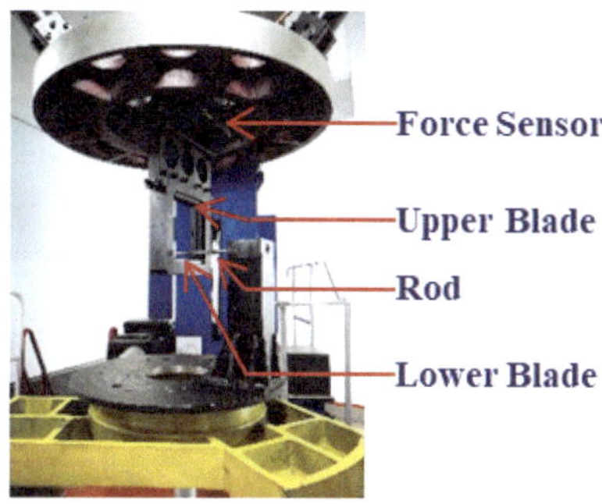

Fig. 9.12 Elastic rod

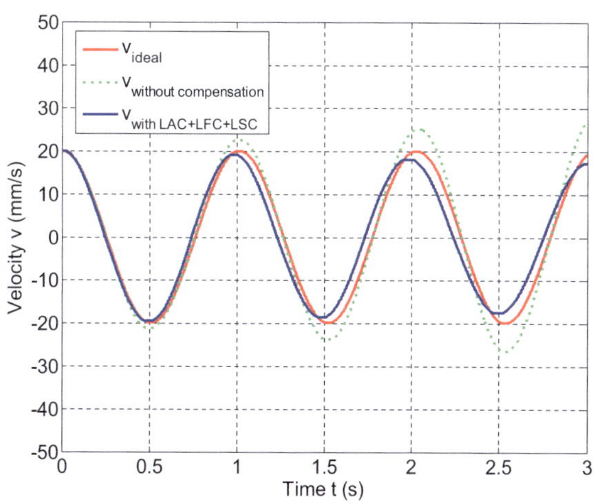

Fig. 9.13 Simulated velocity with LFC + LAC + LSC in the 1 Hz contact experiment on the test device

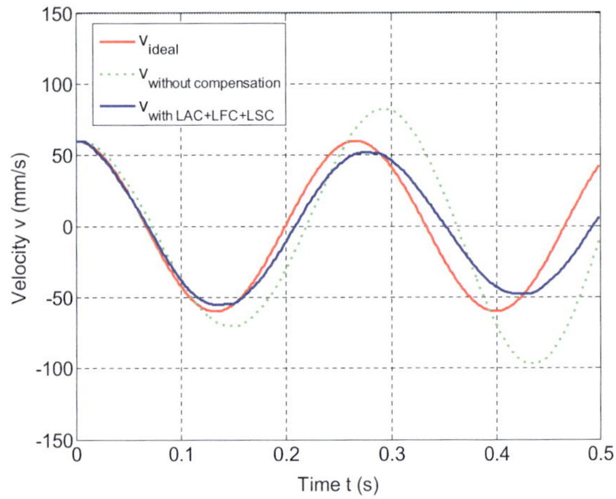

Fig. 9.14 Simulated velocity with LFC + LAC + LSC in the 4 Hz contact experiment on the test device

lower and upper platforms are compensated by a learning-based structure compensator (LSC). The proposed learning-based distortion compensation method does not require the system models. Some verifications using simulations and experiments are given.

References

Hagan, M. T., & Menhaj, M. B. (1994). Training feedforward networks with the Marquardt algorithm. IEEE Transactions on Neural Networks, 5(6), 989–993.

Qi, C. K., Li, D. J., Hu, Y., Zheng, Y., Wang, W. J., Shou, X., & Gao, F. (2023). Learning-based distortion compensation for a hybrid simulator of space docking. IEEE Robotics and Automation Letters, 8(6): 3446–3453.

Chapter 10
Conclusions

The hardware-in-the-loop (HIL) simulation is a very useful and important way to simulate and test the space contact dynamics process of many space missions, such as spacecraft docking, space manipulator operation and so on. It has advantages including the fidelity of physical simulation and the flexibility of numerical simulation. Due to the structure difference between the space docking system and the ground HIL simulation system, it is very difficult to achieve high accuracy of HIL simulation.

An ideal HIL simulator requires that all subsystems have no time delays and errors. However, it is difficult to satisfy these technical requirements using available mechanical and control systems. For a real HIL simulator, there are always time delays or non-ideal dynamics due to the subsystem performance limitations. One challenging issue of the space docking HIL simulator is that these time delays or non-ideal dynamics in the system loops lead to the simulation distortion and even system instability. The HIL simulation distortion often shows the velocity divergence or convergence. This means that the coefficient of restitution (CoR) of velocity (i.e., the ratio of the rebound velocity after the contact and the approach velocity before the contact) is not equal to what occurs in space. The HIL simulation of space docking is often divergent due to the phase delay in the system loops. The convergence distortion can also occur due to dissipative factors, e.g., the structural deformation dynamics of the upper platform.

To better improve the simulation accuracy and guarantee the system stability, the distortion compensation control is a necessary and effective way because it can actively compensate for the time delays and non-ideal dynamics. The main contributions of the book include that the modeling and distortion analysis of the HIL simulation robot are presented, and systematic distortion compensation control methods are proposed.

Chapter 2 introduces the robot system for HIL simulation of space contact dynamics. The principle of the HIL simulation is introduced, and the system structure, mechanical system and control system of the HIL simulation system are presented.

C. Qi et al., *Control of Hardware-In-The-Loop Space Contact Simulation Robot*,
https://doi.org/10.1007/978-981-95-1172-3_10

For the pure time delay of the force measurement system, the first-order phase lead compensation still results in a divergence, while the serial Smith predictor and phase lead compensation achieves a small convergence. The divergence or convergence becomes more obvious as the contact frequency increases. In Chap. 3, the proposed hybrid Smith predictor and phase lead compensation can achieve a little convergence and a higher simulation fidelity. The phase analysis and stability condition are given. The performances of the proposed compensation approaches are tested and compared by the simulations on the elastic contact process.

In Chap. 4, the compensation of the velocity divergence caused by the dynamic response of the lower motion simulator is investigated. The idea is to compensate the contact force error corresponding to the response error of the motion simulator. The compensation force is obtained from the response error of the motion simulator, the contact stiffness and damping. The contact stiffness and damping are identified in real time from the measured contact force, the actual position and velocity. The proposed compensation approach does not require the dynamic response model of the lower motion simulator. The compensation approach is practical and useable when the accurate model of the dynamic response is difficult to obtain due to the time-varying and unknown uncertainties. The effectiveness of the proposed compensation approach is verified by the simulations and experiments.

In Chap. 5, how the structural dynamics of the lower motion simulator affect the accuracy of the HIL simulation is studied. It is found that due to the structural dynamics of the lower motion simulator, divergence and convergence of the HIL simulation are both possible. The conditions of the divergence and convergence of the HIL simulation due to robot structural dynamics of the lower motion simulator are given. The distortion level of the system divergence and convergence is studied. A distortion compensation method for the HIL simulation distortion due to the structural dynamics of the lower motion simulator is proposed. An error analysis and stability analysis after the distortion compensation are given. The software emulations and experiments are used to verify the analysis and the distortion compensation method. Application experiments on real docking mechanisms are also conducted.

In Chap. 6, the simulation convergence problem due to mechanical structural dynamics of the upper platform is investigated. It is found that the phase of the measured docking force between two docking mechanisms is ahead of the desired relative position. To compensate the simulation convergence error, a mechanical structural dynamics model based convergence compensation approach is investigated. The phase analysis and stability analysis of the HIL simulation system are given. The effectiveness of the convergence compensation method are shown in simulations and experiments.

In Chap. 7, considering the force measurement delay, the dynamic response delay of the lower motion simulator and the structural dynamics of the upper platform, a HIL simulation distortion compensation approach is proposed. The force measurement delay is compensated by the phase lead, the dynamic response delay of the lower motion simulator is compensated by the position response error real-time feedback based force compensation, and the structural deformation of the upper platform is compensated by the structural dynamics based force compensation. Both simulation and experimental results show that the approach can compensate the simulation distortion and improve the simulation fidelity.

In Chap. 8, a distributed delay compensation approach is proposed for the HIL simulation system. The force measurement delay, numerical calculation delay, dynamic response delay of the lower motion simulator and structural deformation dynamics of the lower motion simulator are compensated in situ. Compared with lumped compensation, distributed compensation has stronger robustness to the signal noise and better adaptability to the varying contact natural frequency. Simulations show that the proposed compensation approach can guarantee the system stability and simulation accuracy.

In Chap. 9, the distortion compensation problem for a space docking HIL simulator with unknown system loop models is studied. A learning-based distortion compensation method is proposed. The measurement system delay of the contact force is compensated by a learning-based force compensator (LFC). For the motion simulator, the delay of the actuation system of the lower platform is compensated by a learning-based actuation compensator (LAC), and the structural dynamics of the lower and upper platforms are compensated by a learning-based structure compensator (LSC). The proposed learning-based distortion compensation method does not require the system models. Some verifications using simulations and experiments are given.

In order to further improve the simulation accuracy and ensure the stability of the system, future work includes more advanced model-based control of the HIL simulator. When there are uncertainties in the system model of the HIL simulator, advanced artificial intelligence technologies could learn more accurate distortion compensators from experimental data.